2018 EMNLP Workshop SMM4H: 3rd Social Media Mining for Health Applications Workshop & Shared Task

Brussels, Belgium
31 October 2018

ISBN: 978-1-5108-7427-5

EMNLP 2018

Proceedings of the 2018 EMNLP Workshop SMM4H:

3rd Social Media Mining for Health Applications Workshop & Shared Task

October 31, 2018
Brussels, Belgium

Order copies of this and other ACL proceedings from:

Association for Computational Linguistics (ACL)
209 N. Eighth Street
Stroudsburg, PA 18360
USA
Tel: +1-570-476-8006
Fax: +1-570-476-0860
acl@aclweb.org

Preface

Welcome to the 3rd Social Media Mining for Health Applications Workshop and Shared Task - SMM4H.

The total number of users of social media continues to grow worldwide, resulting in the generation of vast amounts of data. With nearly half of adults worldwide and two-thirds of all American adults using social networking, the latest Pew Research Report estimates that 26% of the total users have discussed health information and, of those, 42% have even discussed current medical conditions. Advances in automated NLP and machine learning present the possibility of utilizing this massive data source for biomedical and public health applications, if researchers address the methodological challenges unique to this media.

For its third iteration, the SMM4H workshop takes place in Brussels, Belgium, on November 1, 2018, and is co-located with the 2018 Conference on Empirical Methods in Natural Language Processing (EMNLP). Following on the success of a session and accompanying Workshop on the topic that was hosted at the Pacific Symposium in Biocomputing (PSB) in 2016 and the AMIA Annual Conference in 2017, this workshop aims to provide a forum for the ACL community members to present and discuss NLP advances specific to social media use in the particularly challenging area of health applications, with a special focus given to automatic methods for the collection, extraction, representation, analysis, and validation of social media data for health informatics.

As for the previous years, the workshop includes shared tasks with a particular interest on social media mining for pharmacovigilance. This third execution of the SMM4H shared tasks comprises four subtasks. These subtasks involve annotated user posts from Twitter (tweets) and focus on the (i) automatic classification of tweets mentioning a drug name, (ii) automatic classification of tweets containing reports of first-person medication intake, (iii) automatic classification of tweets presenting self-reports of adverse drug reaction (ADR) detection, and (iv) automatic classification of vaccine behavior mentions in tweets. A total of 14 teams participated and 78 system runs were submitted. Deep learning-based classifiers were the primary approach, but feature-based classifiers and a few ensemble learning systems were also used.

We received very high quality submissions, and present 19 as long and short talks and posters. The organizing committee would like to thank the program committee, consisting of 13 researchers, for their thoughtful input on the submissions, as well as the organizers of EMNLP for their support and management. Finally, a huge thanks to all authors who submitted a paper for the workshop or participated in the shared tasks; this workshop would not have been possible without them and their hard work.

Graciela, Davy, Abeed, Michael

Table of Contents

Conference Program

Wednesday, October 31, 2018

9:00–9:10   *Introduction*
Graciela Gonzalez

9:10–9:30   *Football and Beer - a Social Media Analysis on Twitter in Context of the FIFA Football World Cup 2018*
Roland Roller, Philippe Thomas and Sven Schmeier

9:30–9:50   *Stance-Taking in Topics Extracted from Vaccine-Related Tweets and Discussion Forum Posts*
Maria Skeppstedt, Manfred Stede and Andreas Kerren

9:50–10:10   *Identifying Depression on Reddit: The Effect of Training Data*
Inna Pirina and Çağrı Çöltekin

10:30–11:00   ***Tea Break and Poster Session***

11:00–11:20   *Overview of the Third Social Media Mining for Health (SMM4H) Shared Tasks at EMNLP 2018*
Davy Weissenbacher, Abeed Sarker, Michael J. Paul and Graciela Gonzalez-Hernandez

11:20–11:40   *Changes in Psycholinguistic Attributes of Social Media Users Before, During, and After Self-Reported Influenza Symptoms*
Lucie Flekova, Vasileios Lampos and Ingemar Cox

11:40–12:00   *Thumbs Up and Down: Sentiment Analysis of Medical Online Forums*
Victoria Bobicev and Marina Sokolova

12:00–12:20   *Identification of Emergency Blood Donation Request on Twitter*
Puneet Mathur, Meghna Ayyar, Sahil Chopra, Simra Shahid, Laiba Mehnaz and Rajiv Shah

12:20–12:30   *Dealing with Medication Non-Adherence Expressions in Twitter*
Takeshi Onishi, Davy Weissenbacher, Ari Klein, Karen O'Connor and Graciela Gonzalez-Hernandez

12:30–14:00   ***Lunch***

14:00–14:20   *Detecting Tweets Mentioning Drug Name and Adverse Drug Reaction with Hierarchical Tweet Representation and Multi-Head Self-Attention*
Chuhan Wu, Fangzhao Wu, Junxin Liu, Sixing Wu, Yongfeng Huang and Xing Xie

14:20–14:40   *Classification of Medication-Related Tweets Using Stacked Bidirectional LSTMs with Context-Aware Attention*
Orest Xherija

14:40–15:00   *Shot Or Not: Comparison of NLP Approaches for Vaccination Behaviour Detection*
Aditya Joshi, Xiang Dai, Sarvnaz Karimi, Ross Sparks, Cecile Paris and C Raina MacIntyre

15:00–15:10   *Neural DrugNet*
Nishant Nikhil and Shivansh Mundra

15:10–15:20   *IRISA at SMM4H 2018: Neural Network and Bagging for Tweet Classification*
Anne-Lyse Minard, Christian Raymond and Vincent Claveau

15:20–15:30   *Drug-Use Identification from Tweets with Word and Character N-Grams*
Çağrı Çöltekin and Taraka Rama

15:30–16:00   ***Tea Break and Poster Session***

16:00–16:10   *Automatic Identification of Drugs and Adverse Drug Reaction Related Tweets*
Segun Taofeek Aroyehun and Alexander Gelbukh

16:10–16:20   *UZH@SMM4H: System Descriptions*
Tilia Ellendorff, Joseph Cornelius, Heath Gordon, Nicola Colic and Fabio Rinaldi

16:20–16:30   *Deep Learning for Social Media Health Text Classification*
Santosh Tokala, Vaibhav Gambhir and Animesh Mukherjee

16:30–16:40   *Using PPM for Health Related Text Detection*
Victoria Bobicev, Victoria Lazu and Daniela Istrati

16:40–16:50   *Leveraging Web Based Evidence Gathering for Drug Information Identification from Tweets*
Rupsa Saha, Abir Naskar, Tirthankar Dasgupta and Lipika Dey

16:50–17:00 *CLaC at SMM4H Task 1, 2, and 4*
             Parsa Bagherzadeh, Nadia Sheikh and Sabine Bergler

Football and Beer - a Social Media Analysis on Twitter in Context of the FIFA Football World Cup 2018

Roland Roller, Philippe Thomas, Sven Schmeier
Language Technology Lab, DFKI,
Berlin, Germany
{firstname.surname}@dfki.de

Abstract

In many societies alcohol is a legal and common recreational substance and socially accepted. Alcohol consumption often comes along with social events as it helps people to increase their sociability and to overcome their inhibitions. On the other hand we know that increased alcohol consumption can lead to serious health issues, such as cancer, cardiovascular diseases and diseases of the digestive system, to mention a few. This work examines alcohol consumption during the FIFA Football World Cup 2018, particularly the usage of alcohol related information on Twitter. For this we analyse the tweeting behaviour and show that the tournament strongly increases the interest in beer. Furthermore we show that countries who had to leave the tournament at early stage might have done something good to their fans as the interest in beer decreased again.

1 Introduction

Alcohol can lead to serious health issues. For instance, even though there is no apparent threshold, even one drink of alcohol per day on average can significantly increase the risk of cancer (Roerecke and Rehm, 2012).

Studies have shown, that the exposure to media and commercial communications on alcohol is associated with the likelihood that adolescents will start to drink alcohol, and with increased drinking amongst baseline drinkers (Anderson et al., 2009). In addition to that social events can have a influence on drinking behaviour. In course of this Curtis et al. (2018) apply a Twitter analysis and show that topics such as sporting events, art and food-related festivals are positively correlated to alcohol consumption on US county level.

Various other studies have also explored alcohol-related content on social media, particularly Twitter. Abbar et al. (2015) carry out a food analysis on Twitter and identify weekly periodicities in context of daily volume of tweets mentioning food. Moreover, authors show a correlation between state obesity and caloric value of food (also alcoholic beverages). Instead Culotta (2013) analyse alcohol sales volume in context of Twitter messages. Kershaw et al. (2014) investigate regional alcohol consumption patterns in the UK, while Hossain et al. (2016) explore alcohol consumption patterns in various areas in the US. Curtis et al. (2018) target the prediction of excessive drinking rates and Huang et al. (2017) examine alcohol- and tobacco-related behavourial patterns. And finally, Moreno et al. (2010) carry out a content analysis of adolescents on social media.

This work examines alcohol consumption on Twitter in context of the FIFA Football World Cup 2018. We make use of the results of the above mentioned works, especially the observed correlation between people's behaviour on Twitter and in their real life in context of consumption. The study is carried out across all participating countries of the tournament and explores the influence of the event on the drinking behaviour of people.

2 Experimental Setup

The FIFA World Cup 2018 was taking place from 14th of June until 15th of July. Within the group stage 32 participating teams were playing in 8 groups and completed 3 matches each. After that the two best teams of each group went to the knockout stage. A match usually lasts 90 minutes plus 15 minutes of break ($\approx$2 hours). This work analyses the tweeting behaviours during a match. In the following a match is defined as a time period of one hour before kick-off and three hours after the kick-off. The hour before and after the game are included as supporters might express excitement for the game.

Proceedings of the 3rd Social Media Mining for Health Applications (SMM4H) Workshop & Shared Task, pages 1–4
Brussels, Belgium, October 31, 2018. ©2018 Association for Computational Linguistics

2.1 Data Collection

Tweets over a period between 05/31/2018 and 07/23/2018 were collected, which covers the period of the tournament, but also two weeks before and one week after. As this work analyses messages from all participating countries of the tournament, messages were crawled containing various emojis due to their language independence. The considered emojis are listed in Table 1. In the rest of the work we refer to them as *BEER*, *WINE*, *SAKE* and *BALL*.

Table 1: Emojis used for Twitter crawling

2.2 Country Assignment

In this work, Tweets are examined according to the different participating countries. As only a small number of Tweets contain country related information (8.63%), Tweets lacking this information had to be assigned automatically to the corresponding country of each user. A classifier was trained based on the approach of Thomas and Hennig (2018), which is able to detect the origin of a Tweet based on text and meta information.

As sanity check, Tweets which actually contain information about its origin were compared to the automatically assigned country. On those messages the model achieves an accuracy of above **91%**.

2.3 Preprocessing

Collected Tweets were then mapped to small time intervals of one hour, according to the target label (e.g. beer emoji). For instance a message sent on 06/27/2018 at 5:25 pm (UTC) and containing a beer emoji is assigned to its country, then assigned to the set of beer emoji Tweets, and finally mapped to the time interval 06/27/2018, 5:00 pm. All messages of a particular label and a particular time interval are summed up. In this way a list for each target label is generated containing *time intervals*, *country origin* and *number of relevant Tweets* for this interval. For the following analysis these lists are used as input. One line is considered as TCF (time-country-frequency) triple.

3 Analysis

In the following the different Tweets containing the target emojis are analysed in detail. All ex-

aminations which include significance testing use one-sided paired t-test.

3.1 Outside the Tournament

Figure 1 shows the average number of Tweets per day containing *BEER* emojis before/after the tournament from the participating countries. Participants with less than 50 *BEER* Tweets per day are excluded, to show more meaningful results. We refer to this group as *avg50*. In average, more than half of the *BEER* Tweets per day come from Brazil and England together. Results show no direct correlation to the statistics *Harmful use of alcohol*[1] of the WHO. The reasons for this can only be possibly found out by a longterm analysis going further than pure statistics. In that list Brazil with 7.8 litres of pure alcohol per capita is actually ranked further to the end. Instead countries such as Germany (13.4), France (12.6), England (11.4) and Australia (10.6) would be ranked to the top.

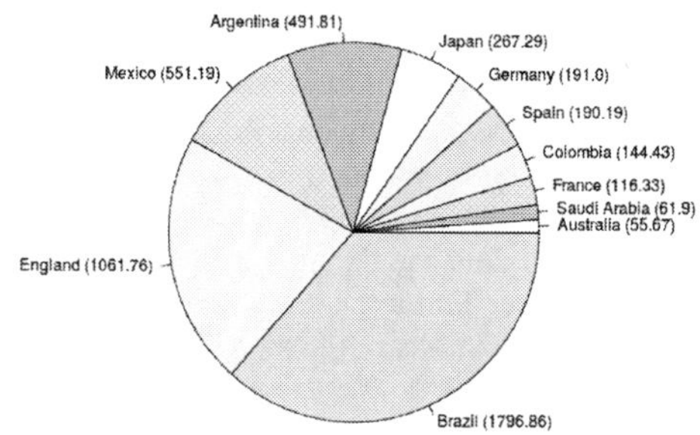

Figure 1: Average number of alcohol related (BEER) Tweets per day outside the tournament

In order to draw a fair comparison to *pure alcohol per capita* the number of active Twitter users must be taken into account. It turns out that dividing the avg. *BEER* Tweets by the number of active Twitter users does not change much. Argentina switches place with England, Saudi Arabia moves to the very end and Japan drops just in front of it. Columbia moves slightly up.

Next mean is calculated for all *BEER* Tweets from *avg50* for each day of the week. The resulting graph is presented in Figure 2 and visualises, similarly to Abbar et al. (2015), particular periodicities. Firstly single days can be recognised as small peaks. Moreover, towards the end of the week, peaks are slightly increased compared to the beginning of the week. Using this data it is for in-

---

[1] http://apps.who.int/gho/data/node. sdg.3-5-viz, accessed 19.07.2018

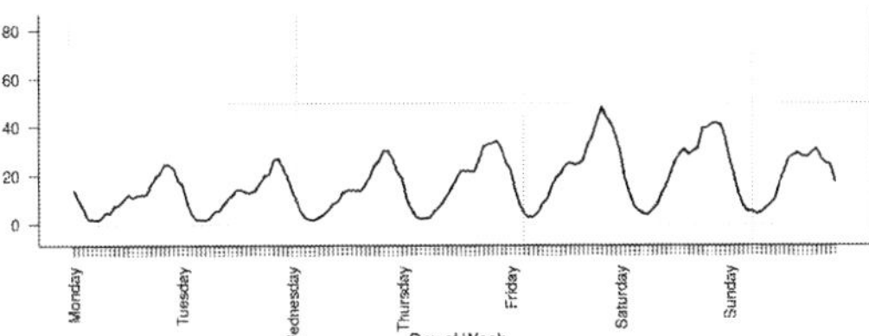

Figure 2: Mean of all alcohol related (*BEER*) Tweets before/after the tournament from *avg50*

stance possible to deduct, that **people tweet significantly more about alcohol on the evening** (from 4pm until 1am) (p<0.001). Moreover, the data also shows, that **people tweet significantly more about alcohol at the weekend** (Friday 4pm - Monday 6am) (p<0.001). in comparison to the rest of the week.

3.2 The Tournament

In this subsection alcohol related Tweets during the tournament are examined. The first question to address is whether supporters of their national team tweet more during the match in comparison to other periods. Reference periods are the days after each match during the same time slots.

The analysis shows that **people from 19 countries use *BEER* significantly more when their team is playing** (p<0.05, 10 of them with p<0.001). Among the 13 other countries, only Japan and Saudi Arabia are from *avg50*. Interestingly Croatia, which reached the final, does not show any significant increase, but the general usage of *BEER* is generally very low here. Considering *WINE* Tweets, only Brazil, Poland and Belgium and for *SAKE* only Mexico show a significant increase in Tweets during the matches of their team (p<0.05). However the number of Tweets are low in comparison to *BEER*.

Figure 3 presents an overview on how the tournament influences the average usage of *BEER* per day of *avg50*, while the team is in the tournament. France shows the largest increase of BEER Tweets per day in average of more than 107%, followed by Japan with 35%. The increase from Japan is surprising as Japanese people do not tweet significantly more during the matches of their team. Possible explanations might be that matches are broadcasted often late in the evening due to the time difference to Russia. For this reason people might meet up earlier, thus start drinking earlier. Another explanation can be just the fact that there is a high interest for the tournament in general in

Japan.

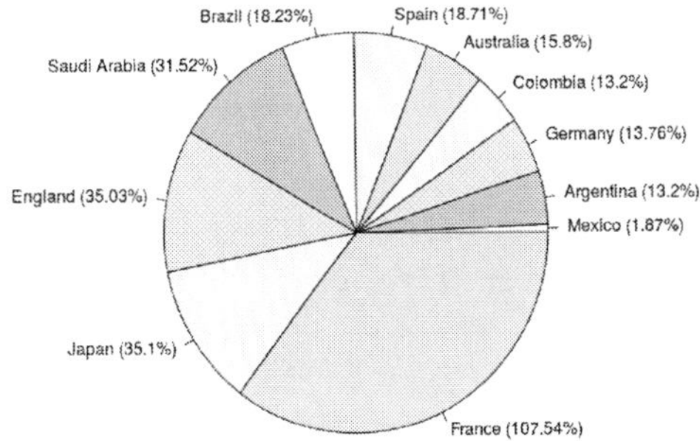

Figure 3: Increase of beer related Tweets per day during World Cup, until leaving the tournament

Brazil instead shows 'only' an increase of 18%. Even with this small increase (in comparison to others) Brazil remains the country with the largest number of avg. *BEER* Tweets per day. Considering all countries Morocco has the strongest increase (407.5%) and also the strongest decrease after leaving the tournament (-68.97%). On the other hand, outside the tournament Morocco has a very low number of *BEER* Tweets per day (0.48), so the increase might be not too serious. Interestingly only Peru shows a decrease during the tournament (-16.4%). Considering the *avg50*, Colombia and Brazil showed the strongest decrease when their team left the tournament with -18.81% and -10.59% respectively.

Generally the results show, that from almost all countries more alcohol related Tweets can be found during the tournament. Moreover, in most cases the avg. number of *BEER* Tweets decreases when the team leaves the tournament. However, in some cases an increase in Tweets can be detected. Senegal for instance increases the number of *BEER* Tweets up to 54%, followed by Uruguay (28%) and Australia (10%).

3.3 Top-5 Matches

This subsection analyses the different matches of the tournament for popularity in terms of *BEER* and *BALL*. In order to have a fair comparison data is normalized first. The average number of Tweets of each country outside the tournament is subtracted from the number of Tweets during the tournament at a given time and day.

Table 2 presents the Top-5 matches involving *BEER* and *BALL*. The table shows that more people use the football emoji than the beer emoji. In terms of *BALL*, the Top-5 list contains the final

(France-Croatia), the opening (Russia-Saudi Arabia) and some other games involving recent European and World Champions.

🍺	#	⚽	#
Mexico-Sweden German-S. Korea	1313	Portugal-Spain	5100
Brazil-Belgium	1305	France-Croatia	4637
Serbia-Brazil Switzerland-C. Rica	1250	Russia-S. Arabia	4253
Nigeria-Iceland	1104	Germany-Mexico	3798
Brazil-Costa Rica	1092	France-Argentina	3616

Table 2: Top-5 matches of the tournament in terms of beer and football emoji (normalized)

From *BEER* perspective we find on the first and third palce matches which took place in parallel. Considering that, the most popular single match was Brazil-Belgium in the Quarter Final. Ranked 4th is Nigeria-Iceland, which is surprising, as both countries are not tweeting much about beer. Analysing the results in more detail reveals, that all 7 matches took place on a day Brazil played. Even though Tweets were normalized, the influence of Brazilian *BEER* Tweets before and after a match of their team is enormous, so that even the Nigeria-Iceland match achieved a high rank.

4   Results

This work presented a short analysis of alcohol related emojis in context of the FIFA football World Cup 2018. With the start of the tournament we showed, that most countries strongly increase the number of Tweets containing beer emojis. As many people tweet less after their national team left the tournament, we draw the conclusion that leaving the tournament early, as Germany did, is the healthiest solution - unless you are Peru. We also showed that people of many participating teams of the tournament tweet significantly more about alcohol during a match of their team. Furthermore we presented the increase of alcohol related Tweets during the tournament and the most popular games in terms of beer and football emojis. Finally we showed, that Brazil tweets by far the most about beer. Cheers!

Acknowledgements

This research was supported by the German Federal Ministry of Economics and Energy (BMWi) through the project MACSS (01MD16011F).

References

Sofiane Abbar, Yelena Mejova, and Ingmar Weber. 2015. You tweet what you eat: Studying food consumption through twitter. In *Proceedings of the 33rd Annual ACM Conference on Human Factors in Computing Systems*, CHI '15, pages 3197–3206, New York, NY, USA. ACM.

Peter Anderson, Avalon de Bruijn, Kathryn Angus, Ross Gordon, and Gerard Hastings. 2009. Impact of alcohol advertising and media exposure on adolescent alcohol use: A systematic review of longitudinal studies. *Alcohol and Alcoholism*, 44(3):229–243.

Aron Culotta. 2013. Lightweight methods to estimate influenza rates and alcohol sales volume from twitter messages. *Language resources and evaluation*, 47(1):217–238.

Brenda Curtis, Salvatore Giorgi, Anneke E. K. Buffone, Lyle H. Ungar, Robert D. Ashford, Jessie Hemmons, Dan Summers, Casey Hamilton, and H. Andrew Schwartz. 2018. Can twitter be used to predict county excessive alcohol consumption rates? *PLOS ONE*, 13(4):1–16.

Nabil Hossain, Tianran Hu, Roghayeh Feizi, Ann Marie White, Jiebo Luo, and Henry A. Kautz. 2016. Precise localization of homes and activities: Detecting drinking-while-tweeting patterns in communities. In *Proceedings of the Tenth International Conference on Web and Social Media, Cologne, Germany, May 17-20, 2016.*, pages 587–590.

Tom Huang, Anas Elghafari, Kunal Relia, and Rumi Chunara. 2017. High-resolution temporal representations of alcohol and tobacco behaviors from social media data. *Proceedings of the ACM on human-computer interaction*, 1(CSCW).

Daniel Kershaw, Matthew Rowe, and Patrick Stacey. 2014. Towards tracking and analysing regional alcohol consumption patterns in the uk through the use of social media. In *Proceedings of the 2014 ACM conference on Web science*, pages 220–228. ACM.

Megan A. Moreno, Leslie R. Briner, Amanda Williams, Libby Brockman, Leslie Walker, and Dimitri A. Christakis. 2010. A content analysis of displayed alcohol references on a social networking web site. *Journal of Adolescent Health*, 47(2):168 – 175.

Michael Roerecke and Jürgen Rehm. 2012. Alcohol intake revisited: Risks and benefits. *Current Atherosclerosis Reports*, 14(6):556–562.

Philippe Thomas and Leonhard Hennig. 2018. Twitter Geolocation Prediction Using Neural Networks. In *Language Technologies for the Challenges of the Digital Age*, pages 248–255, Cham. Springer International Publishing.

Stance-taking in topics extracted from vaccine-related tweets and discussion forum posts

Maria Skeppstedt[1,2], Manfred Stede[2], Andreas Kerren[1]

[1]Department of Computer Science and Media Technology, Linnaeus University, Växjö, Sweden
{maria.skeppstedt,andreas.kerren}@lnu.se
[2]Applied Computational Linguistics, University of Potsdam, Potsdam, Germany
stede@uni-potsdam.de

Abstract

The occurrence of stance-taking towards vaccination was measured in documents extracted by topic modelling from two different corpora, one discussion forum corpus and one tweet corpus. For some of the topics extracted, their most closely associated documents contained a proportion of vaccine stance-taking texts that exceeded the corpus average by a large margin. These extracted document sets would, therefore, form a useful resource in a process for computer-assisted analysis of argumentation on the subject of vaccination.

1   Introduction

Opinions towards vaccination that are expressed in discussion forums and in social media, as well as frequently occurring arguments given in support of these opinions, might help us to better understand reasons behind vaccine hesitancy.

There are previous studies in which such texts have been manually analysed (Grant et al., 2015; Faasse et al., 2016), as well as studies in which topic modelling has been applied for analysing texts about vaccination (Tangherlini et al., 2016; Surian et al., 2016; Skeppstedt et al., 2018).

Through topic modelling, it is possible to automatically extract topics that occur frequently in a text collection. For topic modelling to be a useful strategy for mining text collections for frequently occurring arguments, however, at least some of the topics extracted must correspond to stance positions or arguments given for these positions.

The aim of this study is to investigate if topic modelling is suitable for extracting arguments from two types of document collections that consist of laymen-produced texts about vaccination. We, therefore, measured the occurrence of stance-taking towards vaccination in the documents that were most closely associated with automatically extracted topics from two different corpora.

2   Background

There are previous studies that use topic modelling in computer-assisted processes to find frequently occurring arguments in a document collection (Sobhani et al., 2015; Skeppstedt et al., 2018). Documents that had been manually annotated as not containing argumentation/stance-taking were, however, removed in those two previous studies, i.e., no evaluation of the effects of including neutral documents when performing topic modelling was carried out. For most types of document collections, it is not known beforehand in which documents a stance towards the target of interest is taken or not. Therefore, the setting used here is more widely applicable, i.e., to use topic modelling on an entire text collection, without removing documents in which no stance is taken. In both of these two previous studies, the topic modelling algorithm NMF (Lee and Seung, 2001), i.e., Nonnegative Matrix Factorisation, was shown appropriate for extracting arguments from short argumentative texts. We, therefore, used this algorithm in our experiments.

3   Method

We used topic modelling to automatically extract important topics from two different vaccination corpora, both consisting of English text that predominantly had been written by people without a medical background. We, thereafter, measured the proportion of stance-taking texts among the texts that were most related to these topics, and compared it to the proportion of stance-taking texts in the entire corpus.

3.1   Document collections

As a proxy for texts containing arguments, we used texts in which stance is expressed, since such texts are likely to also contain a motivation for the

Proceedings of the 3rd Social Media Mining for Health Applications (SMM4H) Workshop & Shared Task, pages 5–8
Brussels, Belgium, October 31, 2018. ©2018 Association for Computational Linguistics

position taken. The documents, from each of the two corpora, were divided into two groups based on whether they had been annotated as taking a stance towards vaccination or not, i.e., into the two groups *stance-taking* and *non-stance-taking*.

The first collection consists of posts from discussion threads on the topic of vaccination (Skeppstedt et al., 2017) that contain at least one of the following character combinations: "vacc", "vax", "jab", "immunis", and "immuniz". Posts annotated as taking a stance *for* or *against* vaccination were combined into the group *stance-taking* texts, and posts annotated as *undecided* were assigned the category *non-stance-taking*.

The second collection consists of tweets containing the HPV vaccine-related keywords "HPV", "human papillomavirus", "Gardasil", and "Cervarix" (Du et al., 2017). We combined tweets annotated according to the categories *Positive* and *Negative* to form the category *stance-taking* tweets, and tweets annotated as *Neutral* and *Unrelated* as *non-stance-taking* tweets.

Before applying topic modelling, the following were removed from the texts: standard English stop words, the terms that had been used for gathering the documents, hash tags, user names, URLs and links. Duplicated and near-duplicated documents were also removed from the collections. Documents with identical spans of texts that consisted of more then eight consecutive tokens were counted as near-duplicates. For documents consisting of ten or fewer tokens, a shorter (proportional to the length) cut-off was instead applied for classifying two documents as near-duplicates.

3.2 Applying topic modelling

Separate topic models were constructed for the two document collections, using the NMF class from scikit-learn (Pedregosa et al., 2011). For each topic extracted by the NMF model, the corresponding terms and documents associated with the topic are given as output, as well as their level of association with the topics.

The output of the NMF algorithm is non-deterministic, typically generating slightly different topics when run several times. Therefore, to achieve more reliable results, we followed an approach, for instance used by Baumer et al. (2017), in which the algorithm is re-run several times and only topics that occur in the output from all re-runs are retained. Before checking which topics

occurred in all re-runs, potential outliers were removed from the set of outputs from the re-runs.

We ran the algorithm 100 times with the setting to, for each re-run, return a term set consisting of the 50 terms most closely associated with each of the topics extracted by the algorithm. A topic was counted as stable when there was at least a 70% overlap between the pairs of term sets returned for a topic, for all 90 retained re-runs of the algorithm.

Potential outliers among the outputs were determined by measuring the average term overlap between the re-run outputs. That is, for each re-run, one combined set consisting of all terms associated with all topics from this re-run was constructed. Thereafter, the average overlap between this combined term set and the corresponding sets from the other re-runs was measured, i.e., the combined term sets constructed in the same fashion for each one of the other re-runs. The outputs from the 10% of the re-runs that had the lowest overlap were discarded as potential outliers, and were thus not included when calculating the stability of the extracted topics.

To avoid having to decide on a fixed number of topics in advance, which is normally required from an NMF user, we started by requesting the algorithm to extract 20 topics, and thereafter gradually decreased the number of topics requested until a maximum of 25% of the extracted topics were discarded as non-stable.

4   Results

After the near-duplicate filtering were 1,108 and 2,250 documents retained, for the discussion threads and the tweets, respectively. The proportions of stance-taking documents among the documents that were ranked by the algorithm as the top-*n* documents most typical to the extracted topics are shown in Table 1. These were compared to the 95% confidence interval for the proportion of stance-taking documents among n documents randomly sampled from the corpus.[1] Measurements were carried out for $n=35$ and $n=100$. The method used had yielded 90 re-run outputs, which each one of them contained a slightly different document ranking for the topics extracted. For each of the topics, we therefore extracted the 100 most top-ranked documents for every re-run, and ranked

---

[1]Calculated according to Preston (2012), i.e., 95% of all possible samples of size n yield a proportion $\hat{p}$ within $1.96\sqrt{p\left(1-p\right)/n}$ of the true proportion p.

Topics for discussion threads	Stance top n	
	$n=35$	$n=100$
people/think/mmr/like/child/really	71%	79%
rubella/women/immune/pregnant/girl	**94%**	**88%**
risk/child/disease/risks/carry/catching	82%	**90%**
immunity/herd/checked/wanes	85%	84%
mumps/meningitis/urabe/uk/mmr	88%	83%
000/10/offit/theory/cope/think/paul	74%	75%
children/damaged/unvaccinated	77%	83%
cough/whooping/brother/caught/mum	80%	80%
stance in entire corpus	80%	80%
corpus stance, 95% conf. interval	± 13	± 7.9

Topics for tweets	Stance top n	
	$n=35$	$n=100$
cancer/cervical/cause/prevention	42%	46%
girls/boys/10/need/vaccinated	**74%**	**73%**
vaccine/safe/child/effects/study	60%	**70%**
cancers/caused/related/prevent/protect	57%	45%
rhode/island/graders/mandates/7th	51%	51%
teens/getting/cdc/vaccinations	48%	**56%**
women/men/young/risk/infection/ask	48%	43%
vaccination/rates/low/states/adolescent	**74%**	**65%**
vaccinates/tdap/safety/teen	51%	**57%**
shot/got/doctor/tomorrow/arm	57%	52%
health/public/mandate/dept/activists	40%	41%
love/epidemic/documentary/television	*14%*	*31%*
vax/anti/age/cdc/proven/harm/govt	**80%**	**76%**
stance in entire corpus	44%	44%
corpus stance, 95% conf. interval	± 16	± 10

Table 1: The proportion of stance-taking documents among the top 35 and 100 most typical documents for each extracted topic. Each topic is represented by its most closely associated terms.

these documents according to the sum of the documents topic-association value over the 90 re-runs.

For the figures in Table 1, the stance proportion that lies below the 95% confidence interval for the stance proportions of n randomly selected documents is marked with italics and those that lie above are marked in boldface. That is, the document rankings (top 35 or top 100) that contain a smaller or larger density of stance-taking texts, than had the same number of documents been randomly selected, are marked in italics or boldface.

For discussion forum texts, for which the collection-level proportion of stance-taking was already high, the proportions among the documents extracted for the topics were similar to the document-level proportion. The general trend was a slight increase in stance-taking documents, with one topic that had a stance-taking proportion above the 95% confidence interval for the top-35 documents and two topics that fulfil this criterion for the top-100 topics.

Also for the tweets, a majority of the topics had associated documents with a proportion of stance-taking that did not differ significantly from a random sampling from the document collection. However, some of the topics contained a very high proportion of stance-taking, in comparison to the proportion in the entire document collection. This resulted in that, for the tweets, there was a statistically significant difference for three topics also when extracting only the top 35 most typical documents. These top 35 documents were made up of document sets consisting of semantically coherent tweets. The topic *girls/boys/10...* mainly consisted of posts advocating HPV vaccine for both boys and girls, often also providing the argument that it prevents cancer. The documents belonging to *vax/anti/age/...* typically took the opposite stance, and often contained a questioning of whether there is a proof that HPV vaccination prevents cancer, or warnings against perceived adverse effects of HPV vaccination. The topic *vaccination/rates/low...*, which consisted of expressions of worries about HPV vaccination rates being low, forms an example of that stance-taking does not always imply that arguments are given. That is, although most of the tweets associated with this topic clearly take a stance in favour of vaccination, no direct arguments are given here.

There was also one tweet topic with a very low proportion of stance-taking among its associated documents, that is, the topic *love/epidemic/documentary...* which consisted of many tweets that, in different ways, but in a neutral manner, announced a documentary about HPV.

5   Discussion and conclusion

A typical practical application of the method studied here would be the case in which an analyst aims at finding frequently occurring vaccine-related arguments in a document collection that is too large for a fully manual analysis. The analyst would then instead perform an analysis at which only a subset of the texts would be read, i.e., those automatically extracted through topic modelling.

We have here shown, for the two document

collections investigated, that there are topics extracted which have associated documents that contain a larger proportion of vaccine stance-taking texts than the average document collection. That is, these document sets would form a useful resource for such an analyst who searches for vaccine-related argumentation.

The fact that there might also be topics extracted which do not contain argumentation, i.e., topics similar to the *love/epidemic...* topic, should not pose a large obstacle to the analysis, as long as there are other topics that have associated documents in which stance is taken. That is, at least for the documents extracted for the topic *love/epidemic...*, it is evident after reading only a few documents, that this topic is uninteresting for the task of finding argumentation. Documents closely associated with such topics can, therefore, be excluded from the analysis after a quick inspection. This would enable the analyst to focus on the other topics, which have associated documents that do contain argumentation.

Acknowledgments

The study was funded by the Swedish Research Council (Vetenskapsrådet), through the project "Navigating in streams of opinions: Extracting and visualising arguments in opinionated texts" (No. 2016-06681).

References

Eric P. S. Baumer, David Mimno, Shion Guha, Emily Quan, and Geri K. Gay. 2017. Comparing grounded theory and topic modeling: Extreme divergence or unlikely convergence? *Journal of the Association for Information Science and Technology*, 68(6):1397–1410.

Jingcheng Du, Jun Xu, Hsingyi Song, Xiangyu Liu, and Cui Tao. 2017. Optimization on machine learning based approaches for sentiment analysis on HPV vaccines related tweets. *J Biomed Semantics*, 8(1):9.

Kate Faasse, Casey J. Chatman, and Leslie R. Martin. 2016. A comparison of language use in pro- and anti-vaccination comments in response to a high profile Facebook post. *Vaccine*, 34(47):5808–5814.

Lenny Grant, Bernice L. Hausman, Margaret Cashion, Nicholas Lucchesi, Kelsey Patel, and Jonathan Roberts. 2015. Vaccination persuasion online: A qualitative study of two provaccine and two vaccine-skeptical websites. *J Med Internet Res*, 17(5):e133.

Daniel D. Lee and H. Sebastian Seung. 2001. Algorithms for non-negative matrix factorization. In *Advances in neural information processing systems*, pages 556 – 562.

Fabian Pedregosa, Gaël Varoquaux, Alexandre Gramfort, Vincent Michel, Bertrand Thirion, Olivier Grisel, Mathieu Blondel, Peter Prettenhofer, Ron Weiss, Vincent Dubourg, Jake Vanderplas, Alexandre Passos, David Cournapeau, Matthieu Brucher, Matthieu Perrot, and Edouard Duchesnay. 2011. Scikit-learn: Machine learning in Python. *Journal of Machine Learning Research*, 12:2825–2830.

Scott Preston. 2012. Math 158: Confidence interval for a proportion. http://www.oswego.edu/~srp/158/CI%20Proportion/CI%20for%20a%20 Proportion.pdf.

Maria Skeppstedt, Andreas Kerren, and Manfred Stede. 2017. Automatic detection of stance towards vaccination in online discussion forums. In *Proceedings of the International Workshop on Digital Disease Detection using Social Media 2017 (DDDSM-2017)*, pages 1–8, Stroudsburg, PA, USA. Association for Computational Linguistics.

Maria Skeppstedt, Andreas Kerren, and Manfred Stede. 2018. Vaccine hesitancy in discussion forums: Computer-assisted argument mining with topic models. In *Building Continents of Knowledge in Oceans of Data: The Future of Co-Created eHealth*, number 247 in Studies in Health Technology and Informatics, pages 366–370. IOS Press.

Parinaz Sobhani, Diana Inkpen, and Stan Matwin. 2015. From argumentation mining to stance classification. In *Proceedings of the 2nd Workshop on Argumentation Mining*, Stroudsburg, PA, USA. Association for Computational Linguistics.

Didi Surian, Dat Quoc Nguyen, Georgina Kennedy, Mark Johnson, Enrico Coiera, and Adam G Dunn. 2016. Characterizing Twitter discussions about HPV vaccines using topic modeling and community detection. *J Med Internet Res*, 18(8):e232.

Timothy R. Tangherlini, Vwani Roychowdhury, Beth Glenn, Catherine M. Crespi, Roja Bandari, Akshay Wadia, Misagh Falahi, Ehsan Ebrahimzadeh, and Roshan Bastani. 2016. "Mommy blogs" and the vaccination exemption narrative: Results from a machine-learning approach for story aggregation on parenting social media sites. *JMIR Public Health Surveill*, 2(2):e166.

Identifying depression on Reddit: the effect of training data

Inna Pirina
Department of Linguistics
University of Tübingen
inna.pyrina@gmail.com

Çağrı Çöltekin
Department of Linguistics
University of Tübingen
ccoltekin@sfs.uni-tuebingen.de

Abstract

This paper presents a set of classification experiments for identifying depression in posts gathered from social media platforms. In addition to the data gathered previously by other researchers, we collect additional data from the social media platform Reddit. Our experiments show promising results for identifying depression from social media texts. More importantly, however, we show that the choice of corpora is crucial in identifying depression and can lead to misleading conclusions in case of poor choice of data.

1 Introduction

Clinical depression, also referred to as major depressive disorder, is a serious mental condition that can interfere with normal daily life activities. One of the many risks of clinical depression is suicide – research has indicated that approximately two-thirds of people who die by suicide were dealing with depression at the time of death (Richards and O'Hara, 2014). Meanwhile, according to the World Health Organization, nearly 50% of people with clinical depression worldwide remain untreated. One of the main reasons why the disorder is ignored is believed to be under-diagnosis (Sheenan, 2004).

One of the many ways that the condition could be manifested is in the way people write: the words they choose and the general tone of the produced text are affected by the disorder (Reece et al., 2017). Due to the stigma around clinical depression, people tend to turn to the Internet, thus, making the data gathered from social media a valuable source of literary cues that could help identify depression from texts. Moreover, the ease of obtaining data makes Internet an attractive source for the purpose.

Early research on the relation between language and depression has been mostly theoretical, mainly focusing on the linguistic features that are manifested in 'depressed language', such as negatively-valenced words Beck et al. (1987), and frequent use of first-person pronouns Pyszczynski et al. (1987). These observations have been verified using corpus studies (Rude et al., 2004; Pennebaker et al., 2008), indicating, indeed, certain aspects of linguistic output is related to the speaker's or author's mental state.

A challenge in investigating the link between depression and the linguistic output is obtaining suitable data. And, one of the easy (and fruitful) data source for this purpose has been the Internet, in particular social media platforms (Ramirez-Esparza et al., 2008; Coppersmith et al., 2015).

Most of the earlier works have been focused on analyzing the language used by depressed individuals, and/or finding linguistic correlates of the depression. A more applicable approach to monitoring public or individual mental health requires explicit identification of depression from the linguistic samples. Such an application can complement the conventional diagnosis methods, and, if proven successful, it can be useful for diagnosis where conventional methods are not applicable. Similar to some of the recent studies (Coppersmith et al., 2015; Yates et al., 2017; Lynn et al., 2018), our aim in this paper is identifying depression from linguistic data. Using (mainly) corpora we gather from the social media platform Reddit, we experiment with a number of different classification models. Our focus here is on selection of corpora for reliable and generalizable analysis or identification of depression from the social media data.

2 Methods

2.1 Data

In part of our experiments, we use the data collected by Ramirez-Esparza et al. (2008), which

Proceedings of the 3rd Social Media Mining for Health Applications (SMM4H) Workshop & Shared Task, pages 9–12
Brussels, Belgium, October 31, 2018. ©2018 Association for Computational Linguistics

consists of 400 forum posts by depressed individuals. Ramirez-Esparza et al. (2008) used a similarly sized data set from a breast cancer forum as the 'control group' in their analyses. Since the control data was not available to us, we report results using an alternative set of documents collected by Gorbunova (2017) as the negative class in our classification experiments.

Our data was gathered from Reddit, which hosts over 10 000 online communities (also known as 'subreddits') of anonymous users united by common interests or discussion topics. In all data sets described below, we only collect the original posts, 'submissions', not the comments.

As an approximation to the data collection method of Ramirez-Esparza et al. (2008), we collect data from a relatively large subreddit that is devoted to depression, where authors seek support from the community. Similar to Ramirez-Esparza et al. (2008), we also collect the number of posts from subreddit devoted to breast cancer discussion, as the control set (or negative class). Since the differences between depression and breast cancer may involve serious topical differences, we also collect yet another set of posts from 'family' and 'friendship advice' subreddits, which we presume is topically more similar to depression subreddit.

In all of the cases above, however, the posts in the both positive and negative classes are topically specific. In practice, we would like to identify depression from everyday language, not necessarily language used for talking about depression, and seeking community support. As our more realistic example, we collected a number of posts following a protocol similar to Coppersmith et al. (2015) and Yates et al. (2017). First, we looked for expressions like 'I was just diagnosed with depression', on the depression subreddit. Unlike Yates et al. (2017), we do not manually check the sampled texts. As a result, a certain number of false positives are expected. For each author mentioning a diagnosis, we searched for the postings of the same author within a month of the original post in other subreddits, excluding some potentially related ones like 'Anxiety', 'mentalhealth' and 'depression_help'. The resulting posts are written by authors with (likely) depression, and to a large extend topically different than that of depression subreddit. To keep the training set size the same as the other data sets we use, we randomly sample 400 posts obtained in this manner for training, and

another 400 posts for testing. Another difference from Yates et al. (2017) is that our training and test instances are the documents, not the authors. We also sample randomly the same amount of posts as our texts with authors without depression, from the same set of subreddits, but excluding the authors that posted in the depression subreddit during the time period we used for our investigation. For each setting, we pick only one document for each author.

In sum, we experiment with 8 data sets:

DSF Posts from Depression Support Forums (Ramirez-Esparza et al., 2008)

DND Posts from Non Depression Forums (Gorbunova, 2017)

DS Posts from Depression Support subreddit

BC Posts from Breast Cancer subreddit

FF Posts from subreddits related to Family and Friends

DO Posts from authors with (probable) Depression posted on Other forums

ND Posts from authors with (probably) No Depression

All data sets have 400 training set items, and the data sets DO and ND also have additional 400 posts used as a reasonable test set.

2.2 Classifiers and tuning

We have experimented with a relatively large number of classification methods, including logistic regression and recurrent neural networks, in a number of different settings. In our experiments, the support vector machines (SVMs) with a combination of character and word n-grams of various sizes performed the best. We only report the experiments with SVM models.

In all cases we used linear SVMs with bag-of-n-grams features. SVMs are known to work well in a number of other text classification problems in this setting. The character and word n-grams of various sizes are extracted from the texts, and weighted using BM25 algorithm (Robertson et al., 2009). We optimized maximum order of character and word n-grams as well as the SVM margin parameter C through random search. A 5-fold cross-validation is performed for each parameter setting explored. For each experiment we report the setting where average scores over the 5-fold cross validation is the highest. The BM25 parameters 'k1' and 'b' were not optimized, and set to 0.75 and 2.0 respectively. For experiments with class imbalance, we

Model	5-fold F1	Test set F1
DSF–NDF	94.75	64.05
DS–BC	98.62	56.88
DS–FF	92.25	55.62
DS–ND	91.75	56.48
DO–ND	68.12	67.49
allD–allND	91.40	58.28

Table 1: Best 5-fold CV results obtained on each classification setting, together with the performance of the system on the test set. The model descriptions list data used for positive and negative class respectively. The data sets are explained in Section 2.1. The last row combined all 'positive' and 'negative' data sets, except the DO and ND sets. The scores are percentages.

used class weights during training to overcome the class imbalance problem.

All models were implemented in Python programming language, using scikit-learn package (Pedregosa et al., 2011). The source code for the classification models and data collection scripts are available at `https://github.com/Inusette/Identifying-depression`.

2.3 Evaluation

For evaluating the models, we report the standard measures of F_1 score (harmonic mean of precision and recall). We use the 'binary' version of the scores with positive class being text from authors with depression.

2.4 Experiments and results

We train 6 SVM classifiers, using different data sets descried in Section 2.1. Table 1 presents the performance comparison of the classifier on a number of different settings.

Each row in Table 1 presents F1-score of a binary SVM classifier on the data set as well as the performance of the same system on the test set consisting of DO and ND. Since each model is tuned for F1-score, the precision and recall values are rather balanced, and are not reported in Table 1. In general, 5-fold cross validation results are rather high, especially if both data sets are specific. Best results are obtained when both data sets are very specific, as in DS–BC case. The success of the classifier goes down as the texts belonging to the negative class comes from less specific domains. And in fact, the worse in-dataset results are obtained in our target setting, during which the classification of documents written by authors with diagnosed depres-

sion in non-depression related topics (DO) against the documents on the similar topics written by (presumably) healthy authors (ND). The gap between all other settings and DO–ND setting is rather large.

We also observe a very sharp drop of performance between the 5-fold cross validation results and the results on the test set. Interestingly, the most successful model (except DO–ND) on the test data is the forum data which is expected to be rather different from the all others which came from Reddit.

The last row of Table 1 reports the performance of a model where positive/negative instances of all other (except DO and ND) settings are combined. The resulting model is trained on more data, however, its data sources are not as harmonized as in other settings. As a result, it performs comparably, but worse than other specific models. However, the non-specificity seems to slightly help in the test set, resulting in better than all others (except DSF–NDF setting).

3 Discussion and conclusions

In this paper we reported a number of experiments on detecting depression from language samples collected from social media. Being able to detect depression from linguistic material is interesting both theoretically, and due to its potential applications as a diagnostic aid or for monitoring of public or individual mental health. These goals are viable only if we can identify depression to a successful degree. There has been a number promising results for detecting depression from the writing samples, particularly from social media texts (Ramirez-Esparza et al., 2008; Coppersmith et al., 2015; Reece et al., 2017; Lynn et al., 2018).

In this study, our focus has been the selection of sources for successful detection of depression from social media text. Our results clearly show that careful selection of sources is important for not obtaining illusionary results. This is particularly important if one intends to use the resulting systems in practical applications. However, it may be equally important, not to get wrong conclusions for more theoretically oriented research as well.

Another important contribution of our works is the use of Reddit for the purpose. There has been relatively few studies using Reddit for investigating linguistic aspects of mental health (De Choudhury and De, 2014; Yates et al., 2017). We believe Reddit's emphasis on anonymity is useful for obtaining less biased results. Reddit corpora also has the advantage of availability,[1] which can help reproducing the earlier results. Furthermore, not having length limitation like Twitter, a popular choice in other studies, may also be important in some cases. The F_1-scores we obtained on Reddit corpus, although higher than the results reported in Yates et al. (2017), is lower than the earlier results on Twitter (Coppersmith et al., 2015). This could potentially be due to the small number of training instances in our study. However, further investigation is needed for understanding the differences.

In this study we only reported results from linear classifiers, using simple character and word bag-of-n-gram features. These models are simple, fast, language independent, and performed better than other systems we experimented with, including a number of deep learning architectures (this is in line with some earlier work where same models and methodolgy is used on similar tasks, e.g., Çöltekin and Rama, 2016, 2018). Furthermore, although our focus in this paper has been their performance, the linear models are also more open to analysis, allowing investigation of (types) of features that are useful for the task.

References

Aaron T Beck, A John Rush, Brian F Shaw, and Gary Emery. 1987. *Cognitive Therapy of Depression*. Guilford Press.

Çağrı Çöltekin and Taraka Rama. 2016. Discriminating similar languages with linear svms and neural networks. In *Proceedings of the Third Workshop on NLP for Similar Languages, Varieties and Dialects (VarDial3)*, pages 15–24, Osaka, Japan.

Çağrı Çöltekin and Taraka Rama. 2018. Tübingen-Oslo at SemEval-2018 task 2: SVMs perform better than RNNs at emoji prediction. In *Proceedings of the 12th International Workshop on Semantic Evaluation (SemEval-2018)*, New Orleans, LA, United States.

Glen Coppersmith, Mark Dredze, Craig Harman, Kristy Hollingshead, and Margaret Mitchell. 2015. Clpsych 2015 shared task: Depression and ptsd on twitter. In *Proceedings of the 2nd Workshop on Computational Linguistics and Clinical Psychology: From Linguistic Signal to Clinical Reality*, pages 31–39. Association for Computational Linguistics.

Munmun De Choudhury and Sushovan De. 2014. Mental health discourse on Reddit: Self-disclosure, social support, and anonymity. In *Proceedings of the Eight International AAAI Conference on Weblogs and Social Media (ICWSM)*, Ann Arbor, Michigan, USA.

Anastasia Gorbunova. 2017. Predicting depression from online communication: comparison of three classification techniques. Master's thesis, University of Tübingen, Tübingen, Germany.

Veronica Lynn, Alissa Goodman, Kate Niederhoffer, Kate Loveys, Philip Resnik, and H. Andrew Schwartz. 2018. Clpsych 2018 shared task: Predicting current and future psychological health from childhood essays. In *Proceedings of the Fifth Workshop on Computational Linguistics and Clinical Psychology: From Keyboard to Clinic*, pages 37–46. Association for Computational Linguistics.

Fabian Pedregosa, Gaël Varoquaux, Alexandre Gramfort, Vincent Michel, Bertrand Thirion, Olivier Grisel, Mathieu Blondel, Peter Prettenhofer, Ron Weiss, Vincent Dubourg, Jake Vanderplas, Alexandre Passos, David Cournapeau, Matthieu Brucher, Matthieu Perrot, and Édouard Duchesnay. 2011. Scikit-learn: Machine learning in Python. *Journal of Machine Learning Research*, 12:2825–2830.

James W. Pennebaker, Cindy K. Chung, Ewa Kacewicz, and Nairan Ramirez-esparza. 2008. The psychology of word use in depression forums in English and in Spanish: Testing two text analytic approaches. In *ICWSM*.

Tom Pyszczynski, Kathleen Holt, and Jeff Greenberg. 1987. Depression, self-focused attention, and expectancies for positive and negative future life events for self and others. *Journal of Personality and Social Psychology*, 52(5):994–1001.

Nairan Ramirez-Esparza, Cindy K Chung, Ewa Kacewicz, and James W Pennebaker. 2008. The psychology of word use in depression forums in English and in Spanish: Testing two text analytic approaches. In *International Conference on Weblogs and Social Media*, pages 102–108.

Andrew G. Reece, Andrew J. Reagan, Katharina L. M. Lix, Peter Sheridan Dodds, Christopher M. Danforth, and Ellen J. Langer. 2017. Forecasting the onset and course of mental illness with twitter data. *Scientific Reports*.

C Steven Richards and Michael W O'Hara. 2014. *The Oxford Handbook of Depression and Comorbidity*. Oxford Library of Psychology. Oxford University Press.

Stephen Robertson, Hugo Zaragoza, et al. 2009. The probabilistic relevance framework: BM25 and beyond. *Foundations and Trends® in Information Retrieval*, 3(4):333–389.

Stephanie Rude, Eva-Maria Gortner, and James Pennebaker. 2004. Language use of depressed and depression-vulnerable college students. *Cognition and Emotion*, 18(8):1121–1133.

DV Sheenan. 2004. Depression: underdiagnosed, undertreated, underappreciated. 13.

Andrew Yates, Arman Cohan, and Nazli Goharian. 2017. Depression and self-harm risk assessment in online forums. In *Proceedings of the 2017 Conference on Empirical Methods in Natural Language Processing*, pages 2968–2978, Copenhagen, Denmark. Association for Computational Linguistics.

---

[1]The corpus we use is publicly available at `https://files.pushshift.io/reddit/submissions/`.

Overview of the Third Social Media Mining for Health (SMM4H) Shared Tasks at EMNLP 2018

Davy Weissenbacher[†]**, Abeed Sarker**[†]**, Michael Paul**[‡]**, Graciela Gonzalez-Hernandez**[†]
[†]DBEI, Perelman School of Medicine, University of Pennsylvania, PA, USA
[‡]Information Science University of Colorado Boulder, CO, USA
{dweissen,abeed,gragon}@pennmedicine.upenn.edu
mpaul@colorado.edu

Abstract

The goals of the SMM4H shared tasks are to release annotated social media based health related datasets to the research community, and to compare the performances of natural language processing and machine learning systems on tasks involving these datasets. The third execution of the SMM4H shared tasks, co-hosted with EMNLP-2018, comprised of four subtasks. These subtasks involve annotated user posts from Twitter (tweets) and focus on the (i) automatic classification of tweets mentioning a drug name, (ii) automatic classification of tweets containing reports of first-person medication intake, (iii) automatic classification of tweets presenting self-reports of adverse drug reaction (ADR) detection, and (iv) automatic classification of vaccine behavior mentions in tweets. A total of 14 teams participated and 78 system runs were submitted (23 for task 1, 20 for task 2, 18 for task 3, 17 for task 4).

1 Introduction

The third execution of the SMM4H shared tasks built on the success of the two previous shared task workshops, which were held at the Pacific Symposium on Biocomputing (PSB) in 2016 and at the AMIA Annual Symposium in 2017. In line with the previous shared tasks, the data comprised of medication mentioning posts from Twitter, which were retrieved using the Twitter public streaming API. For this iteration, We designed and provided annotated data for four tasks. The annotated data were made publicly available for download. The performances of participating systems were compared on blind evaluation sets for each task.

1.1 Shared Task Design

Teams were allowed to participate in one or multiple tasks. In order to analyze cross-task application of classification techniques, all the tasks

for this year's execution focused on text classification. Manually annotated training data for the four tasks were made available to the participants in May, 2018. Unlabeled evaluation data was released in July, 2018. Evaluations of participant submissions were conducted from 29th July to 2nd of August. In total, 14 teams participated in the shared tasks and 78 system runs were accepted from them (maximum of three submissions per team per task). We received 23 submissions for task 1, 20 for task 2, 18 for task 3, 17 for subtask 4. Participating teams were invited to submit system descriptions to describe their approaches to the tasks. We provide descriptions of the four tasks and the associated data in the following sections/subsections.

2 Task Descriptions

2.1 Tasks

The primary goal of the SMM4H shared tasks is to promote community driven development and evaluations of systems focusing on social media based health data. This year's tasks involved medication-mentioning user posts from Twitter. We included two tasks from the last execution at AMIA and two new task. Outlines of the tasks are as follows:

1. Automatic classification of posts mentioning a drug name. In this binary classification task, the systems were required to distinguish tweets mentioning any drug names or dietary supplement. Often run first in automatic pipelines mining health related information in social media, the performances obtained on this task conditions the overall performances of the entire pipelines and their usefulness. This proposed task was new and intended to establish common baselines for future research.

2. Automatic classification of medication intake mentioning posts. This is a three-class

Proceedings of the 3rd Social Media Mining for Health Applications (SMM4H) Workshop & Shared Task, pages 13–16
Brussels, Belgium, October 31, 2018. ©2018 Association for Computational Linguistics

text classification task. Each medication-mentioning tweet is categorized into three classes—definite intake (where the user presents clear evidence of personal consumption), possible intake (where it is likely that the user consumed the medication, but the evidence is unclear), and no intake (where there is no evidence that the user consumed the medication).

3. Automatic classification of ADR mentioning tweets. This is a binary text classification task for which systems were required to predict if a tweet mentions an ADR or not. Such a system is crucial for active surveillance of ADRs from social media data as most of the medication-related chatter in the domain does not represent ADRs.

4. Automatic classification of vaccine behavior mentions in tweets. Specifically, English-language tweets are classified to indicate whether the user intends to receive a seasonal influenza (flu) vaccine (Huang et al., 2017). It is a binary classification task where the positive class indicates that the user has received or intends to receive the current flu vaccine, and all other tweets (which are filtered with vaccine-related keywords) are labeled negative. Such a classifier can be used to measure patterns in vaccination behaviors across populations.

To facilitate the shared task, we made available large annotated Twitter data sets. The overall shared task was designed to capitalize on the interest in social media mining and appeal to a diverse set of researchers working on distinct topics such as natural language processing, biomedical informatics, and machine learning. The different subtasks presented a number of interesting challenges including the noisy nature of the data, the informal language of the user posts, misspellings, and data imbalance. We provide details of the data used for each of the four above-mentioned tasks, and the tasks themselves, in the following subsections.

2.2 Data

The dataset made available for the shared tasks were collected from Twitter using the public streaming API. Task 1 and task 4 included new and unpublished annotated datasets provided as training and testing sets. Tasks 2 and 3 re-used existing

training datasets from the SMM4H-2017 shared tasks where the SMM4H-2017 shared tasks' evaluation sets were included in the training datasets used this year. These datasets had been made available with our prior publication following the execution of the past workshop (Sarker et al., 2018).

Task 1: Drug names detection. Participants were given tweets with binary annotation, indicating the presence or absence in the tweet of one or more drug names/dietary supplement, manually created. The data were released in two phases. An initial set of 9,622 tweets were made available[1] for training to any participants. The test set composed of 5,382 tweets was distributed only to registered participants. Both training and test sets were balanced with 4647/2530 tweets mentioning no drug and 4975/2852 tweets mentioning at least one drug, respectively. All participants were evaluated using common metrics for binary classification: Precision, Recall and F-score for tweets mentioning a drug.

Task 2: Medication Intake Classification. Participants were provided with tweets that have been manually categorized into three classes: definite intake, possible intake and no intake. Data was released in the same manner as task 1. 17,773 annotated tweets were made available for training. The evaluation set consisted of 5000 tweets. For this task, the evaluation metric was micro-averaged F-score for the definite intake and possible intake classes. This metric was chosen for evaluation because the tweets belonging to these two classes are of interest in social media based drug safety surveillance systems, while the no intake class primarily represents noise.

Task 3: ADR Classification. Participants were provided with the training/development set containing tweets which were annotated in a binary fashion to indicate the presence or absence of ADRs. A total of 25,633 annotated tweets were made available for training. The evaluation set consisted of 5000 tweets. The evaluation metric for this task was the F-score for the ADR class, since the primary intent of this task is to be able to filter out ADR indicating tweets from large amounts of noise.

Task 4: Vaccine Behavior Classification. Par-

---

[1] Due to Twitter's privacy policy, the actual tweets were not shared publicly. We made available the TweetIDs and UserIDs for the tweets as well as a download script to download all publicly available tweets.

Team	Institution(s)-Country	P	R	F
ART	Tata Consultancy Services Limited, India	0.785	0.880	0.830
CIC-NLP	Instituto Politecnico Nacional, Mexico	0.920	0.899	0.910
ClaC	Concordia University, Canada	0.788	0.769	0.778
IIT_KGP	Indian Institute of Technology, India	0.918	0.840	0.877
IRISA	INRIA-IRISA, France	0.922	0.906	0.914
LILU	Technical University of Moldova, Moldova	0.841	0.860	0.850
Techno	University Abou Bekr Belkaid, Algeria	0.905	0.855	0.879
THU_NGN	Tsinghua University, China	0.933	0.904	**0.918**
Tub-Oslo	University of Tubingen, Germany University of Oslo, Norway	0.917	**0.907**	0.912
UChicagoCompLx	University of Chicago, USA	**0.937**	0.891	0.914
UZH	University of Zurich, Switzerland	0.927	0.878	0.902

Table 1: System performances for each team for task 1 of the shared task. Precision, Recall and F-score over the drug mention class is shown. Top score in each column is shown in bold.

Team	Institution(s)-Country	P	R	F
ClaC	Concordia University, Canada	0.402	0.366	0.383
IIT_KGP	Indian Institute of Technology, India	0.408	0.407	0.408
IRISA	INRIA-IRISA, France	0.434	0.501	0.465
LIGHT	Indian Institute of Technology, India	0.520	0.491	0.505
Techno	University Abou Bekr Belkaid, Algeria	0.327	0.432	0.372
Tub-Oslo	University of Tubingen, Germany University of Oslo, Norway	0.478	0.458	0.468
UChicagoCompLx	University of Chicago, USA	**0.654**	**0.783**	**0.713**
UZH	University of Zurich, Switzerland	0.371	0.437	0.401

Table 2: System performances for each team for task 2 of the shared task. Micro-averaged Precision, Recall and F-score over the definite intake and possible intake classes are shown. Top score in each column is shown in bold.

ticipants were provided with two sets of annotated data, one with 8,181 tweets and the other with 1,665 tweets, where approximately one third of the tweets are labeled positive. Tweets were annotated with binary labels indicating whether the user intends to receive a flu vaccine. The evaluation set consisted of 161 tweets. The evaluation metric for this task was the F-score for the positive class, since the primary intent of this task is to identify if someone has received a flu vaccine.

3 Results

Task 1: Fourteen teams registered to participate in the task and 23 submissions from eleven teams were included in the final evaluations. Table 1 presents the performances of the best systems for each teams having submitted. Team THU_NGN had the best performing system for this task, obtaining a F-score of 0.9182.

Task 2: Eight teams submitted twenty system runs for the final evaluations. Table 2 presents the performances of the best systems for each team in terms of micro-averaged F-score for the intake and possible intake classes. UChicagoCompLx achieved top spot with a micro-averaged F-score of 0.71.

Task 3: Nine teams submitted eighteen system runs for the final evaluations. Table 3 presents the performances of the best systems for each team in terms of ADR class F-score. Team THU_NGN obtained the best F-score of 0.522.

Task 4: Nine teams submitted seventeen system runs for the final evaluations. Table 4 presents the performances of the best systems for each team. Team CARRDS obtained the best F-score of 0.887.

4 Conclusion

The submitted systems employed a wide range of deep learning based classifiers but also feature-based classifiers and few attempts with ensemble learning systems. The system descriptions

Team	Institution(s)-Country	P	R	F
ART	Tata Consultancy Services Limited, India	0.332	0.547	0.413
CIC-NLP	Instituto Politecnico Nacional, Mexico	0.314	0.529	0.394
IIT_KGP	Indian Institute of Technology, India	0.189	0.643	0.292
IRISA	INRIA-IRISA, France	0.378	**0.649**	0.478
Techno	University Abou Bekr Belkaid, Algeria	0.434	0.344	0.383
THU_NGN	Tsinghua University, China	0.442	0.636	**0.522**
Tub-Oslo	University of Tubingen, Germany University of Oslo, Norway	**0.638**	0.317	0.424
UChicagoCompLx	University of Chicago, USA	0.370	0.464	0.411
UZH	University of Zurich, Switzerland	0.455	0.436	0.445

Table 3: System performances for each team for task 3 of the shared task. Precision, Recall and F-score over the ADR class are shown. Top score in each column is shown in bold.

Team	Institution(s)-Country	P	R	F
CARRDS	CSIRO-Data61, Australia	**0.918**	0.859	**0.887**
ClaC	Concordia University, Canada	0.700	0.897	0.787
IRISA	INRIA-IRISA, France	0.867	0.833	0.850
IIT_KGP	Indian Institute of Technology, India	0.800	0.769	0.784
LIGHT	Indian Institute of Technology, India	0.824	0.897	0.859
LILU	Technical University of Moldova, Moldova	0.829	0.808	0.818
techno	University Abou Bekr Belkaid, Algeria	0.870	0.859	0.865
Tub-Oslo	University of Tubingen, Germany University of Oslo, Norway	0.840	0.872	0.855
UChicagoCompLx	University of Chicago, USA	0.791	**0.923**	0.852

Table 4: System performances for each team for task 4 of the shared task. Precision, Recall and F-score over the positive class is shown. Top score in each column is shown in bold.

that have been published with the shared task proceedings provide further details about these methods and the relative performances of each. The successful execution of the shared tasks suggests that this is an effective model for encouraging community-driven development of systems for social media based heath related text mining, and warrants further future efforts.

References

Xiaolei Huang, Michael C. Smith, Michael J. Paul, Dmytro Ryzhkov, Sandra C. Quinn, David A. Broniatowski, and Mark Dredze. 2017. Examining patterns of influenza vaccination in social media. In *AAAI Joint Workshop on Health Intelligence*.

Abeed Sarker, Maksim Belousov, Jasper Friedrichs, Kai Hakala, Svetlana Kiritchenko, Farrokh Mehryary, Sifei Han, Tung Tran, Anthony Rios, Ramakanth Kavuluru, Berry de Bruijn, Filip Ginter, Debanjan Mahata, Saif M. Mohammad, Goran Nenadic, and Graciela Gonzalez-Hernandez. 2018. Data and systems for medication-related text classification and concept normalization from twitter: Insights from the social media mining for health (smm4h) 2017 shared task. *Journal of the American Medical Informatics Association*, article in press; doi:10.1093/jamia/ocy114.

Changes in psycholinguistic attributes of social media users before, during, and after self-reported influenza symptoms

Lucie Flekova[1]*, Vasileios Lampos[2], Ingemar J. Cox[2]†

[1] Amazon Alexa AI, Aachen, Germany
[2] Department of Computer Science, University College London, UK
lflekova@amazon.de, {v.lampos, i.cox}@ucl.ac.uk

Abstract

Previous research has linked psychological and social variables to physical health. At the same time, psychological and social variables have been successfully predicted from the language used by individuals in social media. In this paper, we conduct an initial exploratory study linking these two areas. Using the social media platform of Twitter, we identify users self-reporting symptoms that are descriptive of influenza-like illness (ILI). We analyze the tweets of those users in the periods before, during, and after the reported symptoms, exploring emotional, cognitive, and structural components of language. We observe a post-ILI increase in social activity and cognitive processes, possibly supporting previous offline findings linking more active social activities and stronger cognitive coping skills to a better immune status.

1 Introduction

Stylistic variation in spoken and written communication of different users can provide rich information about them, such as their sociodemographic background (Rao et al., 2010; Argamon et al., 2009; Lampos et al., 2014; Preoţiuc-Pietro et al., 2015; Flekova et al., 2016), personality (Schwartz et al., 2013), mental health (De Choudhury et al., 2013), mood, beliefs, fears or cognitive patterns (Snowdon et al., 1996). At the same time, researchers have been observing relations between factors such as mental health, mental states, personality, happiness, and physical health, including direct relation between individual stress level and resistance to infectious diseases (Cohen and Williamson, 1991; Martin et al., 1995; Friedman, 2000; Smith and Gallo, 2001; Kiecolt-Glaser

et al., 1998; Uchino, 2006). In this paper, we conduct an initial exploratory study linking these two research areas. Using the social media platform of Twitter, we identify users self-reporting symptoms that are descriptive of influenza-like illness (ILI). We analyze the tweets of those users in the periods before, during, and after the reported ILI symptoms, and extract linguistic variables linked to affective, cognitive, perceptual and social processes, as well as personal concerns. We observe a post-ILI increase in social activity and cognitive processes, possibly supporting previous findings that individuals, who spend less time in social activities or are less capable of coping with stress, are associated with a poorer immune status (Friedman, 2000; Pressman et al., 2005; Jaremka et al., 2013; Pennebaker et al., 1997).

2 Related work

Socially stable individuals are at significantly lower risk for disease (Cohen and Williamson, 1991; Martin et al., 1995; Kiecolt-Glaser et al., 1998; Friedman, 2000). Associations were found between personality and likelihood of physical limitations. Chronic negative emotions are associated with suppressed immune functioning, and optimism with lower ambulatory blood pressure and better immune functioning (Smith and Gallo, 2001). Smolderen et al. (2007) examined stress, negative mood, negative affectivity and social inhibition related to increased vulnerability to influenza on participants. They concluded that negative affectivity and perceived stress were associated with higher self-reporting of ILI.

There is considerable evidence that social isolation is associated with poorer health. Those with more types of relationships and those who spend more time in social activities are at lower risk for disease and mortality than their more isolated

---

* Project carried out during the research fellowship at the University College London, prior to joining Amazon

† Also with Department of Computer Science, University of Copenhagen, Denmark

Proceedings of the 3rd Social Media Mining for Health Applications (SMM4H) Workshop & Shared Task, pages 17–21
Brussels, Belgium, October 31, 2018. ©2018 Association for Computational Linguistics

counterparts (Friedman, 2000). Subjectively perceived loneliness and small social networks have also been associated with poorer immune status, greater psychological stress and poorer sleep quality (Pressman et al., 2005; Jaremka et al., 2013). Loneliness was also associated with greater psychological stress and negative affect, less positive affect, poorer sleep efficiency and quality, and elevations in circulating levels of cortisol (Pressman et al., 2005).

Some of these psychological and social variables have been previously successfully identified through an automated stylistic analysis of written text. For example, a series of natural language processing (NLP) workshops has been focusing on predicting depression on Twitter (Coppersmith et al., 2015b,a; Preoţiuc-Pietro et al., 2015), finding that the frequencies of functional words, auxiliary verbs, conjunctions, words indicating cognitive mechanisms, hedging expressions and exclusion words are a strongly predictive feature combination to separate depressed and healthy users. Earlier work on this topic found that authors with depressive tendency are more self-focused, use more frequently the "I" pronoun (Rude et al., 2004), and discuss in social media topics around feelings and sadness (Schwartz et al., 2014).

3   Dataset collection

We randomly sampled 14 million UK tweets, collected in the years 2014-2016, and searched for a small set of word patterns potentially indicative of having the flu based on previous work (Lampos and Cristianini, 2010, 2012; Lampos et al., 2015), such as any combination of {*I have, I feel, I've got*} and {*flu, sore throat, high fever, stupid fever, hate fever, ill*} excluding {*http, rt, jab, shot, you, he, she*}. We obtained 2,600 tweets matching the pattern, which we then manually examined, finally obtaining 1,235 referring to the users themselves being sick with a flu, cold, sore throat, or fever. The false positive tweets were often discussing news about flu, flu vaccination, or social media trends such as (Justin) *Bieber fever* or *cabin fever*.

The 1,235 tweets come from 285 users. These users have been rather verbose on Twitter, producing 7.2 million tweets, responses and retweets over the three years. We decided to monitor the period from 7 days before the user first mentions being sick, to 14 days after this mention, as our first assumption was that the ILI symptoms last about a week since the first tweet. However, after the manual empirical exploration of user tweets over time, we reassessed this hypothesis, and for the rest of this study we are assuming the peak ILI period (i.e., the time period when the flu has the most extreme symptoms) is occurring slightly sooner, i.e. between one day before and two days after the time when a user is self-reporting that has the disease (TSR, time of self-report). We obtained 144,837 tweets, and after filtering out retweets this averaged to 231 tweets per user over these three weeks.

4   Statistical analysis method

We extract textual features using the Linguistic Inquiry and Word Count (LIWC) (Pennebaker et al., 2001), which consists of dozens of lexicons related to psychological processes (e.g., Affective, Cognitive, Biological), personal concerns (e.g., Work, Leisure, Money) and other categories such as Fillers, Disfluencies or Swear words. For each word category, we count a relative occurrence of the words of that category as a proportion to all words for a given user in a given time period.

Per set of days $d \geq 3$, we calculated the mean $\langle o \rangle_d$ of the occurrences o for a single feature as $\langle o \rangle_d = \sum_{i=0}^{N} o_i / N$ with N being the number of users tweeting in that relative period d (e.g. "7 days before TSR" to "5 days before TSR") and o_i being a feature value for one user in that period (e.g. relative proportion of words from category Family to all words tweeted by that user in that period). An example is demonstrated on Figure 2. For each data point $\langle o \rangle_d$, the period d is illustrated with the horizontal bar and the standard error the mean $SE_{\langle o \rangle_d} = \frac{s}{\sqrt{N}}$ as a vertical bar. We then calculate the significance that the mean of the feature two and more days before the ILI symptoms TSR differs from the mean of the feature in the assumed ILI symptom peak interval (one day before to two days after TSR), as well as the significance that the mean of the feature three and more days after the ILI symptoms TSR differs from it. The significance σ is calculated as:

$$\sigma_{\text{before}} = \frac{\langle o \rangle_{\text{before}} - \langle o \rangle_{\text{during}}}{\sqrt{SE^2_{\langle o \rangle_{\text{before}}} + SE^2_{\langle o \rangle_{\text{during}}}}} \tag{1}$$

$$\sigma_{\text{after}} = \frac{\langle o \rangle_{\text{during}} - \langle o \rangle_{\text{after}}}{\sqrt{SE^2_{\langle o \rangle_{\text{during}}} + SE^2_{\langle o \rangle_{\text{after}}}}} \tag{2}$$

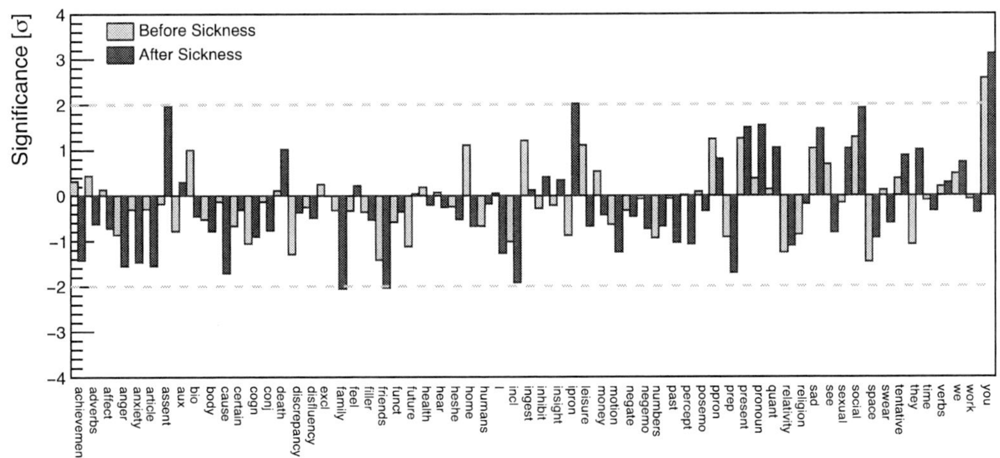

Figure 1: Summary of significance of differences in feature value distributions before and after the self-reported sickness. Generally, we are looking for σ values larger than 1.96, corresponding approximately to two-tailed significance test of $p = 0.05$ for the feature values during and before/after the ILI peak. If in reality the feature distribution in both period was the same, we would observe these or larger differences in $<5\%$ of the cases.

Both the σ and the corresponding p-values are listed in the individual feature plots. Generally, we are looking for σ values larger than 1.96, corresponding to two-tailed significance test of $p = 0.05$, indicating that if in reality the feature distribution in both periods was the same, we would observe these or larger differences in $< 5\%$ of the cases. Both the sigmas and the corresponding p-values are listed in the individual feature plots.

5 Results

Figure 1 shows all differences in feature value distributions before and after the ILI TSR. While the values before typically tend to resemble the values during sickness more closely, the values from day +3 onwards show several significant ($> 2\sigma$, $p < 0.05$) differences. The relative frequency of words from Friends and Family LIWC categories increases after the ILI peak (Fig. 2b and 2a). This indicates that users probably get more socially involved after recovery, however, the relatively low values in the period before the ILI may support the hypothesis that individuals spending less time in social activities are associated with a poorer immune status (Friedman, 2000; Pressman et al., 2005; Jaremka et al., 2013). We also observe a post-ILI increase in the usage of causal words. Causal words connote terms to explain cause and effect (e.g. *reason, why, because*). Increased use of words in this category has been previously found to be related to improved physical health due to stronger coping skills (Pennebaker et al., 1997). There is a 2σ decrease in impersonal pronouns after the ILI peak (Fig. 2e). Higher levels of impersonal pronoun usage have been previously associated to increased anxiety levels (Coppersmith et al., 2015a), hence this change could indicate a post-illness drop in anxiety.

Additional effects observed are: (a) a 2σ post-ILI decrease in assent words such as *agree, OK, yes* (Fig. 2c), which surprisingly signals decreased levels of agreement with the social group and is linked to lower-quality social relationships (Tausczik and Pennebaker, 2010), and (b) a 3σ increase in second person pronouns (Fig. 2f) during the illness, suggesting focus on others.

We found no significant difference in emotion levels or the levels of usage of the pronoun *"I"*.

6 Conclusions and future work

We conduct an initial exploratory study of psycholinguistic attributes of Twitter users before, during, and after self-reported influenza symptoms. We observe a post-ILI change in expressions that correlate with elevated levels of social activity and cognitive processes. Interestingly, instead of an expected increase in using first-person pronoun *"I"* during the ILI peak (focus on self), we observe a significant increase in second person pronouns (*"you"*). We plan to extend this study by including additional 11,000 ILI events.

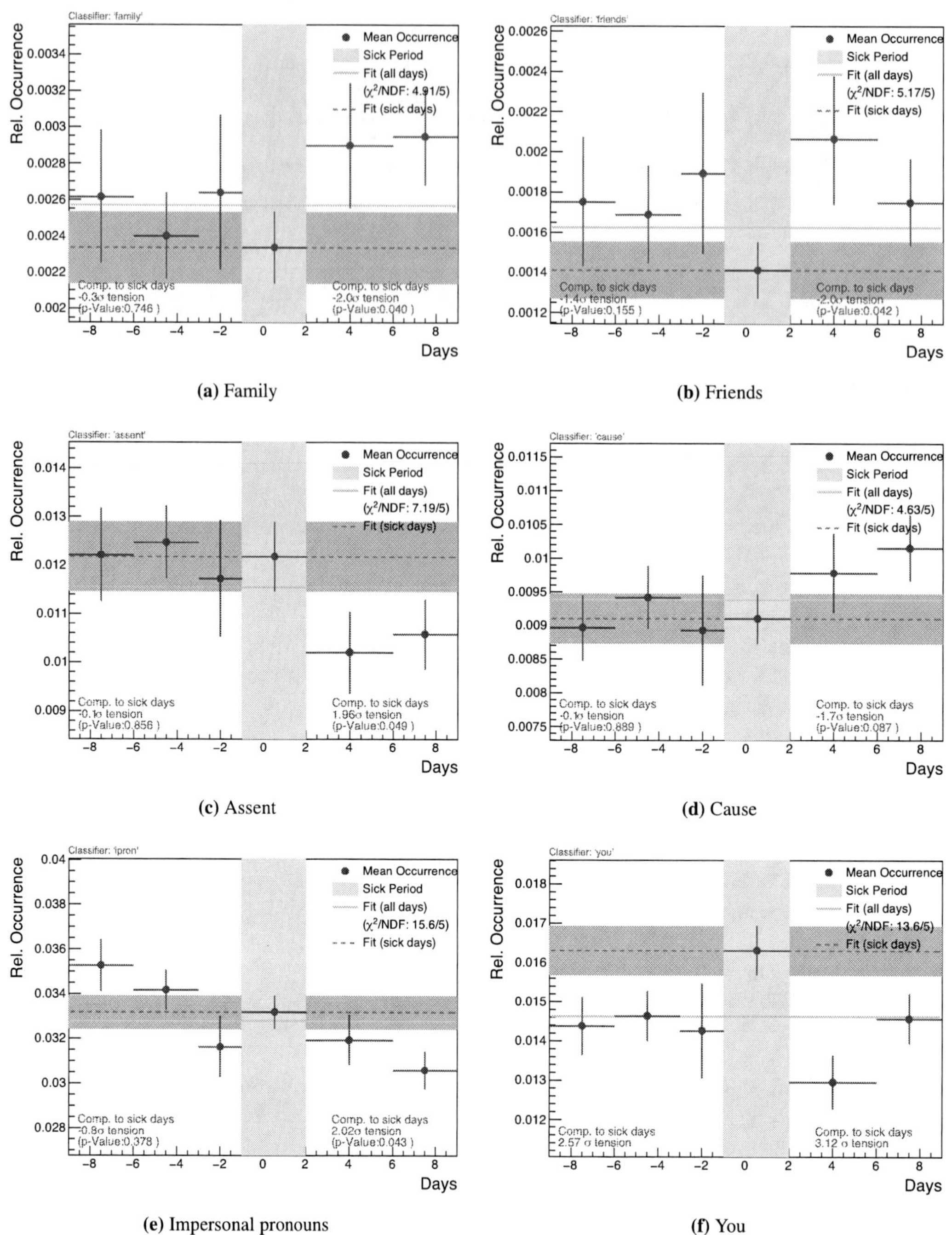

(a) Family

(b) Friends

(c) Assent

(d) Cause

(e) Impersonal pronouns

(f) You

Figure 2: Feature values for selected LIWC categories (specified in the sub-captions) averaged across all diagnosed users for day intervals relative to their ILI symptoms. For each local mean value (blue point), the period of the mean is illustrated with the horizontal bar and the standard error of the mean as a vertical bar. The horizontal blue stripe visually aids to compare to the ILI peak standard error interval, and the vertical grey stripe to the ILI peak period. In addition, an average feature value during the ILI peak is illustrated by a dashed line, compared to the overall average of the feature (yellow line).

Acknowledgements

This work has been supported by the grant EP/K031953/1 (EPSRC).

References

S. Argamon, M. Koppel, J. W. Pennebaker, and J. Schler. 2009. Automatically Profiling the Author of an Anonymous Text. *CACM*, 52(2).

S. Cohen and G. M. Williamson. 1991. Stress and infectious disease in humans. *Psychol. Bull.*, 109(1):5.

G. Coppersmith, M. Dredze, C. Harman, and K. Hollingshead. 2015a. From ADHD to SAD: Analyzing the language of mental health on Twitter through self-reported diagnoses. CLPSYCH '15, pages 1–10.

G. Coppersmith, M. Dredze, C. Harman, K. Hollingshead, and M. Mitchell. 2015b. CLPsych 2015 shared task: Depression and PTSD on Twitter. CLPSYCH '15, pages 31–39.

M. De Choudhury, M. Gamon, S. Counts, and E. Horvitz. 2013. Predicting Depression via Social Media. ICWSM '13, pages 128–137.

L. Flekova, D. Preoţiuc-Pietro, and L. Ungar. 2016. Exploring stylistic variation with age and income on twitter. In *ACL '16*, pages 313–319.

H. S. Friedman. 2000. Long-term relations of personality and health: Dynamisms, mechanisms, tropisms. *J. Pers.*, 68(6):1089–1107.

L. M. Jaremka et al. 2013. Loneliness predicts pain, depression, and fatigue: understanding the role of immune dysregulation. *Psychoneuroendocrinology*, 38(8):1310–1317.

J. K. Kiecolt-Glaser, R. Glaser, J. T. Cacioppo, and W. B. Malarkey. 1998. Marital Stress: Immunologic, Neuroendocrine, and Autonomic Correlates. *Ann. N. Y. Acad. Sci.*, 840(1):656–663.

V. Lampos and N. Cristianini. 2010. Tracking the flu pandemic by monitoring the Social Web. In *Proc. of the 2nd International Workshop on Cognitive Information Processing*, pages 411–416.

V. Lampos and N. Cristianini. 2012. Nowcasting Events from the Social Web with Statistical Learning. *ACM Trans. Intell. Syst. Technol.*, 3(4):1–22.

V. Lampos, E. Yom-Tov, R. Pebody, and I. J. Cox. 2015. Assessing the impact of a health intervention via user-generated Internet content. *Data Min. Knowl. Discov.*, 29(5):1434–1457.

V. Lampos et al. 2014. Predicting and Characterising User Impact on Twitter. EACL '14, pages 405–413.

L. R. Martin et al. 1995. An archival prospective study of mental health and longevity. *Health Psychol.*, 14(5):381–387.

J. W. Pennebaker, M. E. Francis, and R. J. Booth. 2001. *Linguistic Inquiry and Word Count*. Lawerence Erlbaum Associates.

J. W. Pennebaker, T. J. Mayne, and M. E. Francis. 1997. Linguistic predictors of adaptive bereavement. *J. Pers. Soc. Psychol.*, 72(4):863–871.

D. Preoţiuc-Pietro, V. Lampos, and N. Aletras. 2015. An Analysis of the User Occupational Class through Twitter Content. ACL '15, pages 1754–1764.

D. Preoţiuc-Pietro et al. 2015. The role of personality, age, and gender in tweeting about mental illness. CLPSYCH '15, pages 21–30.

S. D. Pressman et al. 2005. Loneliness, social network size, and immune response to influenza vaccination in college freshmen. *Health Psychol.*, 24(3):297–306.

D. Rao, D. Yarowsky, A. Shreevats, and M. Gupta. 2010. Classifying Latent User Attributes in Twitter. SMUC '10.

S. Rude, E.-M. Gortner, and J. Pennebaker. 2004. Language use of depressed and depression-vulnerable college students. *Cogn. Emot.*, 18(8):1121–1133.

H. A. Schwartz et al. 2013. Personality, Gender, and Age in the Language of Social Media: The Open-Vocabulary Approach. *PLoS ONE*, 8(9):e73791.

H. A. Schwartz et al. 2014. Towards assessing changes in degree of depression through Facebook. CLPSYCH '14, pages 118–125.

T. W. Smith and L. C. Gallo. 2001. Personality traits as risk factors for physical illness. *Hand. Health Psychol.*, pages 139–172.

K. G. Smolderen, A. J. Vingerhoets, M. A. Croon, and J. Denollet. 2007. Personality, psychological stress, and self-reported influenza symptomatology. *BMC Public Health*, 7(339).

D. Snowdon et al. 1996. Linguistic Ability in Early Life and Cognitive Function and Alzheimer's Disease in Late Life: Findings from the Nun Study. *JAMA*, 275(7):528–532.

Y. R. Tausczik and J. W. Pennebaker. 2010. The psychological meaning of words: LIWC and computerized text analysis methods. *J. Lang. Soc. Psychol.*, 29(1):24–54.

B. N. Uchino. 2006. Social support and health: a review of physiological processes potentially underlying links to disease outcomes. *J. Behav. Med.*, 29(4):377–387.

Thumbs Up *and* Down: Sentiment Analysis of Medical Online Forums

Victoria Bobicev
Technical University of Moldova

victoria.bobicev@ia.utm.md

Marina Sokolova
IBDA@Dalhousie University and
University of Ottawa

sokolova@uottawa.ca

Abstract

In the current study, we apply multi-class and multi-label sentence classification to sentiment analysis of online medical forums. We aim to identify major health issues discussed in online social media and the types of sentiments those issues evoke. We use ontology of personal health information for Information Extraction and apply Machine Learning methods in automated recognition of the expressed sentiments.

1   Introduction

Computational Health. Online social media became an invaluable and ever growing source of Computational Health (Collier et al, 2017; Sarker et al, 2015). Personal health information, i.e. information about health that individuals share in clinical settings, had been found on Twitter, other social networks, in blogs and medical forums (Sokolova and Schramm, 2011). A diverse language and a subjective style of social media messages stipulate two principal components of Computational Health: i) automated recognition of medical concepts, ii) automated identification of sentiments. The former is essential for extraction of health information (Limsopatham and Collier, 2016); the latter enables to recognize personal attitude in discussion of one's health (Sokolova and Bobicev, 2013).

We apply multi-class and multi-labeled sentence classification in sentiment analysis of online medical forums. We aim to identify major health issues discussed in online social media and the types of sentiments those issues evoke. In order to do this, we adapt ontology of personal health information used in social media studies (Sokolova and Schramm, 2011). By using Machine Learning methods in multi-class classification, we significantly improve over the majority class baseline (paired t-test for all the eight labels: P = 0.0062) and over the look-up results (paired t-test over all the labels, P=0.0208).

2   Related Work.

Sentiment analysis of user-written content has been performed intensely for studies of goods and services reviews, tweets and blogs (Serrano-Guerrero et al., 2015). Khan et al (2016) have shown that a rule-based sentiment classification can be a viable method of sentence-based sentiment analysis. We differentiate between lexicon-based and aspect-based approaches in sentiment analysis studies. The lexicon-based analysis relies on retrieval of lexical expressions of sentiments (Taboada et al, 2011), whereas the aspect-based analysis focuses on sentiments and opinions related to specific features of the product or service (Liu, 2012).

Sentiment analysis of health information is an expanding research domain (Denecke and Deng, 2015). It had been shown that sentiments can be conclusively connected with health issues (Chen and Sokolova, 2018). Health-related texts often express complex sentiments, hence benefit from a multi-label approach in sentiment classification (Bobicev and Sokolova, 2017).

Navindgi et al. (2016) used syntactic features to compare document-level and sentence-level multi-class sentiment classification of online medical forums. They opine that adding social components can benefit the classification results.

Many health-related studies use Twitter data, a popular sphere of public communications (Grover et al, 2018). Tweets had been used in Information Extraction of personal health information (Sokolova et al, 2013), as well as in health studies of specific population groups (Bravo and Goetz, 2017) and in analysis of particular health-related issues (Abbasi et al, 2018). Sokolova et al (2013) had shown that personal pronouns and family relations significantly im-

22

Proceedings of the 3rd Social Media Mining for Health Applications (SMM4H) Workshop & Shared Task, pages 22–26
Brussels, Belgium, October 31, 2018. ©2018 Association for Computational Linguistics

proved accuracy of health information extraction from Twitter.

3 The Data Set Construction

We work with texts harvested from *in vitro fertilization* forums, namely, ivf.ca, with posts annotated by multiple sentiments.[1] The posts are comparatively informative, containing approx. 100-150 words each. Many posts express more than one sentiment and discuss more than one topic. The posts had been studied in a multi-label sentiment classification (Bobicev and Sokolova, 2017). In the said study, multi-label classification has been applied to a complete post, thus leaving aside a nuanced analysis of the expressed sentiments. In the current work, we use *sentences* as the units of the study to gain more detailed information about expressed sentiments.

Sentiment categories. We use two categories *encouragement*, *confusion*, and *facts* introduced in previous studies (Sokolova and Bobicev, 2013).

Encouragement indicates sentiments expressed towards the interlocutors of the post author. The expressed sentiments aim to support and inspire other people reading the posts. At the same time, this support is expressed by describing details of treatment such as: clinics, doctors, procedures or medicines that could lead to the desired outcome.

Confusion generalizes various nuances of negative sentiments: uncertainty, hopeless, frustration, complaint, etc. While analyzing the posts marked by *confusion*, we aim to extract the cause of these negative sentiments; here we differentiate between health issues *per se* and issues of treatment.

Facts is used to label the objective discussions. In posts labeled by *facts* we seek to extract information related to health (e.g., treatment, procedures, prescribed medications).

Health issue categories. The health-related ontology introduced in (Sokolova and Schramm, 2011) was the main resource of Information Extraction procedures. The ontology has been created to study user-written online messages on health-related topics. It contained four main health issue categories: (1) 'Person' with subclasses 'Anatomical parts' and 'Physiological

functions'; (2) 'Health-Related Problems' with subclasses 'Symptoms' and 'Diseases'; (3) 'Health Care System' with subclasses 'Health Care Providers', 'Health Care Setting' 'Health Care Procedures'; (4) Health-Related Environmental Factors.

We expanded the ontology with two new categories: Intakes and External Factors. Our initial version of the ontology listed the following categories: (1) Body: parts, organs, elements, functions; (2) Health conditions: symptoms, diseases; (3) Health care: providers, settings; (4) Health care actions: diagnostics, procedures; (5) Intakes: medicines, supplements, food; (6) External factors: family, work, finances.

However, a simple lookup resulted in high precision and low recall (Precision=0.97, Recall=0.23). The low Recall was due to various spelling of health related terms, especially multi-syllable medical terms (e.g. echocardiography') and specific abbreviations (e.g., ultrasound was written as US or U/S). Unlike in studies of Twitter data (Sokolova et al, 2013), adding personal pronouns and family relations did not improve accuracy of the health information retrieval. In our data, the authors used personal pronouns indiscriminately in description of health issues and other topics. When creating unigram models for posts with health information and without it, we observed that 'I' is the most frequent word in both. The next most frequent pronoun in health related text is 'my' and in non-health related texts - 'you'; family relationship mentioning is actually more frequent in non-health related texts.

The final set of the ontology term categories (i.e., health issues) was as following: (1) Body parts, organs; (2) Health conditions: symptoms, diseases; (3) Health care providers; (4) Actions: procedures; (5) Intakes.

Sentence annotation. We selected 160 posts for sentence annotation and further evaluation by machine learning methods. The selected posts i) had to have 2 or more sentiment labels, ii) had to be an average length (300 - 600 characters, or 50-100 words). Those posts had been split into sentences. Each sentence was manually annotated using two sets of labels: sentiments and health issues mentioned in this sentence.

It is important to note that sentences could have more than one label from the same category, e.g., *encouragement* and *facts, providers* and *organs.* Some sentences had multiple labels and some sen-

---

tences had zero labels. For example, "*So it's a matter of getting the balance right.*" did not have assigned labels, whereas "*I just want to make it clear to anyone with DOR or LOR that there still is hope!*" has been assigned with **encouragement** and **symptoms**.

The annotation resulted in 1087 sentences annotated with the total of 985 labels (Table 1).

Labels	0	1	2	3	4	5
Sentences	490	297	226	61	12	1

Table 1: The statistics on the label distribution.

Further, we worked with the label distribution presented in Table 2.

Sentiments	Health Issues
facts : 213	procedures : 234
encouragement : 110	symptoms : 127
confusion : 70	providers : 86
	organs : 84
	intakes : 61

Table 2: Label distribution in the data set.

4 Empirical Studies

Feature selection. We tokenized each sentence and built the unigram model of the data. All the tokens have been used as features in the initial feature set.

To obtain the best set of features for each label we used Information Gain (IG) to calculate coefficients of the token importance for the current label: $IG\,(token, label) = H(label) - H(label/token)$.

For example, the highest coefficients for the topic 'organs' were: *eggs* - 0.079, *tubes*- 0.021, *egg*- 0.018, *ovary* 0.017, *ovaries* 0.014, and the lowest for the selected features were: *abdominal* - 0.0034, *sorta* - 0034, *like* - 0030. We calculated the coefficients for every word and selected words with the coefficients > 0.

Multi-class classification. We calculated the baseline F-measure (B) where all instances are attributed to the majority class. Thus, F-measure is quite high due to the data imbalance.

To assess difficulty of the multi-class classification, we used a straight-forward look-up to identify each label. The threshold for the label has been selected by balancing Precision (Pr) and Recall (R) of this label recognition. Table 3 shows Pr, R and F-measure (F) calculated for each label.

For the five health issue labels, the look-up non-significantly improved F-measure over the baseline (paired t-test for the five health issue labels: P=0.1308); classification improvement did not happen for the three sentiment labels, albeit F-measure decrease was not significant (paired t-test for the three sentiment labels: P=0.1060).

We used Machine Learning experiments to improve sentence-based sentiment classification. To find algorithms that can improve on the baseline, we applied applied Naïve Bayes, Support Vector Machine, K-Nearest Neighbors, and Decision Tree classifiers from WEKA[2] toolkit.

We applied 10-fold cross validation on the set of the annotated sentences. Table 4 reports the best results for each label. SVM and KNN substantially outperformed other algorithms. The results show that the best results significantly improved over the baseline results: paired t-test for all the labels: P = 0.0062. Improvement over the look-up results is also statistically significant: paired t-test over all the labels, P=0.0208.

However, these experiments treated every label individually and did not reveal relationship among them. To seek relationship among the label categories and the individual labels, we involved multi-label classification.

Label	B	Pr	R	F
facts	0.717	0.735	0.653	0.692
encourage	0.851	0.971	0.600	0.742
confusion	0.904	0.891	0.700	0.784
procedures	0.690	0.759	0.739	0.749
symptoms	0.828	0.958	0.888	0.922
organs	0.887	0.985	0.807	0.887
providers	0.883	0.925	0.860	0.892
intakes	0.917	0.919	0.934	0.927

Table 3: Multi-class labels' lookup results.

Multi-label classification. In multi-label classification (Sorower, 2010), we focused on joint detection of the sentiment and health issues labels assigned to a sentence. We had 667 sentences with at least one label. To convert from a multi-label to a uni-label problem, we used Binary Relevance (BR) problem transformation method. It creates k datasets, each for every single label, and trains the classifier on each of these data sets.

---

[2] https://www.cs.waikato.ac.nz/ml/weka/

Using the lookup method we obtained 292 sentences with Exact Match (EM) = 0.438.

$$ExactMatch = \frac{1}{n}\sum_{i=1}^{n} I(Y_i = Z_i)$$

where n denotes the number of sentences in the

Label	Alg.	B	Pr	R	F
facts	SVM	0.717	0.845	0.854	0.831
encourage.	KNN	0.851	0.897	0.905	0.869
confusion	KNN	0.904	0.947	0.952	0.942
procedures	SVM	0.690	0.848	0.854	0.835
symptoms	SVM	0.828	0.906	0.908	0.884
organs	KNN	0.887	0.954	0.951	0.940
providers	SVM	0.883	0.922	0.930	0.907
intakes	SVM	0.917	0.967	0.966	0.959

Table 4: The best multi-class results of ML algorithms.

data set, Y_i, Z_i are sets of predicted and true labels for sentence i respectively.

EM is the ultimate assessment of accuracy, as it counts only sentences with every label found and identified correctly. This means that the system detected correctly all the labels for more than 40% of sentences (443 labels in total).

The look-up classified 219 sentences with a partial match, where 294 labels were matched correctly, 145 labels were false negative and 115 labels were false positive. 'Match' indicates manually annotated a label found by the lookup; 'false positive' shows that a label was found by the lookup but not by the manual annotation; 'false negative' indicates an annotated label missed by the lookup.

Among 156 completely mismatched sentences, 103 labels were classified as false negative and 96 labels were classified as false positive.

We have applied multi-label Machine Learning algorithms from MEKA toolkit[3]. As in multi-class-classification, we used 10-fold cross-validation. In this task, SVM and Naïve Bayes outperformed the other algorithms. SVM obtained EM = 0.513, F (by label) = 0.438. Naïve Bayes obtained EM = 0.421, F (by label) = 0.406.

The best EM, obtained by SVM, is higher than EM = 0.450 reported for studies of the complete posts (Bobicev and Sokolova, 2017). In addition to classifying a bigger unit, the cited work analyzed only four sentiment labels, whereas we obtained a higher EM in a more complex classification of three sentiment labels and five health issue labels. However, our data set is considerably smaller that the data used in the previous study: 597 sentences vs 1321 posts.

Error analysis. We categorized reasons for errors as follows: (1) linguistic challenges: irony, misspellings, ambiguous sentence structure that requires application of specialized linguistic methods; (2) limitations of the knowledge source, i.e., deficiency of terms in the applied ontology; (3) system limitations, e.g., inability of our system to capture long distance relations of terms and sentiments.

5 Conclusions and Future Work

We present a preliminary sentence-level sentiment analysis of posts gathered from a medical forum. The posts were informative enough to express several sentiments and cover several health issues. As a result, we analyzed a multi-labeled data set, where some labels revealed sentiments and other labels indicated underlying health issues.

We adapted ontology that was previously used in personal health information extraction from a heterogeneous social media data to identify health issues in the data set. Respectively, we added Intake terms and populated the ontology with domain specific terms of In Vitro Fertilization and their slang spellings used by the online forum participants. By using Machine Learning methods in multi-class classification, we have obtained significant improvement over the majority class baseline (paired t-test for all the eight labels: P = 0.0062) and significant improvement over the look-up results (paired t-test over all the labels, P=0.0208). The obtained results on multi-label classification are less conclusive, in part, because a small data set.

Hence, we want to expand the data set through annotation of more posts on the sentence level. This will allow us to use syntactic structures of sentences in order to better capture their semantics.

At the same time, more work should be done for development of an automated and robust system that can reliably classify sentiments and related to them health issues on social media. To improve on Information Extraction, we plan to augment the current ontology.

Finally, we want to test the same approach on posts collected from other medical forums.

---

[3] http://waikato.github.io/meka/

Acknowledgements

We thank the SMM4H anonymous reviewers for thorough and helpful comments.

References

Abbasi, Rabeeh Ayaz, Onaiza Maqbool, Mubashar Mushtaq, Naif R. Aljohani, Ali Daud, Jalal S. Alowibdi, and Basit Shahzad. 2018. Saving lives using social media: Analysis of the role of twitter for personal blood donation requests and dissemination. *Telematics and Informatics* 35(4), pp. 892-912.

Bobicev, Victoria, and Marina Sokolova. 2017. Confused and Thankful: multi-label sentiment classification of health forums. *Proceeding of Canadian Conference on Artificial Intelligence 2017,* pp 284-289.

Bravo, Caroline, and Laurie Hoffman-Goetz. 2017. Social media and men's health: a content analysis of Twitter conversations during the 2013 Movember campaigns in the United States, Canada, and the United Kingdom. *American journal of men's health* 11 (6), pp. 1627-1641.

Chen, Qufei, and Marina Sokolova. 2018. Word2Vec and Doc2Vec in Unsupervised Sentiment Analysis of Clinical Discharge Summaries. *arXiv preprint* arXiv:1805.00352.

Collier, Nigel, Nut Limsopatham, Aron Culotta, Mike Conway, Ingemar J. Cox and Vasileios Lampos. 2017. WSDM 2017 *Workshop on Mining Online Health Reports.*

Denecke, Kerstin, and Yihan Deng. 2015. Sentiment analysis in medical settings: New opportunities and challenges. *Artificial intelligence in medicine,* 64(1):17–27.

Grover, Purva, Arpan Kumar Kar, and Gareth Davies. 2018. "Technology enabled Health"–Insights from twitter analytics with a socio-technical perspective." *International Journal of Information Management* ,43, pp. 85-97.

Khan, Jawad, Byeong Soo Jeong, Young-Koo Lee, and Aftab Alam. 2016. "Sentiment analysis at sentence level for heterogeneous datasets." In *Proceedings of the Sixth International Conference on Emerging Databases: Technologies, Applications, and Theory*, pp. 159-163.

Limsopatham, Nut, and Nigel Collier. 2016. Normalizing medical concepts in social media texts by learning semantic representation. *In Meeting of the Association for Computational Linguistics pp.* 1014-1023.

Liu, Bing. 2012. Sentiment Analysis and Opinion Mining. *Synthesis Lectures on Human Language Technologies*. Morgan & Claypool Publishers.

Navindgi, Amit, Caroline Brun, Cecile Boulard, Scott Nowson. 2016. Steps Toward Automatic Understanding of the Function of Affective Language in Support Groups. *Proceedings of The Fourth International Workshop on Natural Language Processing for Social Media*, pp. 26-33.

Sarker, Abeed, Rachel Ginn, Azadeh Nikfarjam, Karen O'Connor, Karen Smith, Swetha Jayaraman, Tejaswi Upadhaya, and Graciela Gonzalez. 2015. Utilizing social media data for pharmacovigilance: a review. *Journal of biomedical informatics* 54: 202-212.

Serrano-Guerrero, Jesus, Jose A. Olivas, Francisco P. Romero, and Enrique Herrera-Viedma. 2015. Sentiment analysis: A review and comparative analysis of web services. *Information Sciences* 311: pp. 18-38.

Sokolova, Marina, and David Schramm. 2011. Building a Patient-based Ontology for User-written Web Messages. *RANLP 2011.*

Sokolova, Marina, and Victoria Bobicev. 2013. What Sentiments Can Be Found in Medical Forums? *RANLP 2013*, pp. 633–639.

Sokolova, Marina, Stan Matwin, Yasser Jafer, David Schramm. 2013. How Joe and Jane Tweet about Their Health: Mining for Personal Health Information on Twitter. *RANLP 2013.*

Sorower, Mohammad S. 2010. A literature survey on algorithms for multi-label learning. *Technical report, Oregon State University, Corvallis.*

Taboada, Maite, Julian Brooke, Milan Tofiloski, Kimberly Voll, Manfred Stede. 2011. Lexicon-based methods for sentiment analysis. *Computational linguistics*, 37 (2), pp. 267-307

Identification of Emergency Blood Donation Request on Twitter

Puneet Mathur[1], **Meghna Ayyar**[2], **Sahil Chopra**[3],
Simra Shahid[4], **Laiba Mehnaz** [4], and **Rajiv Ratn Shah**[2]

[1]Netaji Subhas Institute of Technology, NSIT-Delhi
`pmathur3k6@gmail.com`
[2]Indraprastha Institute of Information Technology, IIIT-Delhi
`{meghnaa, rajivratn}@iiitd.ac.in`
[3]Maharaja Surajmal Institute of Technology, MSIT-Delhi
`sahilc.msit@gmail.com`
[4]Delhi Technolological University, DTU-Delhi
`{simrashahid_bt2k16, laibamehnaz}@dtu.ac.in`

Abstract

Social media-based text mining in healthcare has received special attention in recent times due to the enhanced accessibility of social media sites like Twitter. The increasing trend of spreading important information in distress can help patients reach out to prospective blood donors in a time bound manner. However such manual efforts are mostly inefficient due to the limited network of a user. In a novel step to solve this problem, we present an annotated Emergency Blood Donation Request (EBDR) dataset[1] to classify tweets referring to the necessity of urgent blood donation requirement. Additionally, we also present an automated feature-based SVM classification technique that can help selective EBDR tweets reach relevant personals as well as medical authorities. Our experiments also present a quantitative evidence that linguistic along with handcrafted heuristics can act as the most representative set of signals this task with an accuracy of 97.89%.

1 Introduction

Sufficiency in the availability of blood in emergency situations can dramatically improve the life expectancy and quality of lives of patients in chronic medical conditions. However, many patients still suffer due to the dual challenges of timely availability and shortfall of required whole blood and blood components. In the case of countries with low rates of blood donation record, blood donation is largely dependent on the families and friends of patients, usually through the word of mouth or peer to peer networking. With the increasing accessibility of social media websites, several instances have emerged where the friends and the family of the patients in need of a blood transfusion have tried to voice their urgent need of blood donation through social media channels. They have reached out to the online community through tweets, Facebook posts, and status updates on popular social media platforms. The effect of such tweets in an emergency situation is largely limited to a user's first few degrees of connections. Thus, it fails to reach the desired donor within the stipulated critical time.

Problem Statement: *Emergency Blood Donation Request (EBDR)* detection is the task of identifying tweets that *explicitly or implicitly mention a necessity for an urgent blood donor.*

1.1 Challenges

The key challenges in preparing the corpus are:

1. **More than one topic categorization:** The task of EBDR tweets prediction is not limited to segregation into binary classes on the basis of certain keywords (like *urgent need, blood required*, etc.). It involves identification of multiple textual modalities such as blood group, quantity of blood required, the disease being treated and presence of personal details for authentication.

2. **Multiple instances of retweets:** It is difficult to obtain a unique set of EBDR tweets as many of the instances of such tweets extracted through the search API are retweets by immediate connections of the original tweet author. In many cases, the tweets describing the same events are spread by rephrasing, morphing or editing the original tweets causing duplication of tweet instances.

1.2 Contributions

The main contributions can be summarized as:

---

[1]The dataset and code are available for research purposes at `https://github.com/pmathur5k10/EBDR`

Proceedings of the 3rd Social Media Mining for Health Applications (SMM4H) Workshop & Shared Task, pages 27–31
Brussels, Belgium, October 31, 2018. ©2018 Association for Computational Linguistics

- Building a corpus of annotated emergency blood donation request tweets divided into three separate datasets using different tweet extraction methodologies.

- Extraction of ancillary handcrafted features from tweets pertaining to the specifications of the requested blood donation.

- Feature modeling using four independent sets of tweet features: linguistic, handcrafted, user specific metadata and textual metadata for the purpose of tweet classification followed by determination of the most relevant set of auxiliary features for SVM based classification.

2 Related Work

Several successful attempts in health text mining have shown that social media can act as a rich source of information for public health monitoring (Broniatowski et al., 2015). MA and Eldredge developed an annotated dataset from consumer drug review posts on social media. Twitter data has been used previously for identifying mentions of medication intake (Mahata et al., 2018a,b), monitoring prescription drug abuse (Hanson et al., 2013). The domain of health text mining also extends to include mental health. Past work on identifying hateful behaviour (Mathur et al., 2018a,b) and suicidal behavior on Twitter (Sawhney et al., 2018a).

3 Dataset

3.1 Dataset Creation

The tweets were collected between 10 May 2018 to 10 July 2018 using Twitter Streaming API. The major problem of extracting tweets for solicitation of blood donation was the infrequent and sporadic nature of their occurrence. Due to the limitation of time restriction imposed on the search query based retrieval, two parallel strategies were developed to build three separate datasets:

1. **Personal Donation Requests(PDR) Dataset** (1311 EBDR, 1511 non-EBDR): A curated list of 53 medical phrases was extracted from selected online blood donation information portals such as American Red Cross[2], Australian Red Cross[3] and NHS Blood and transplant[4] by using TF/IDF (Ramos et al.,

2003) to identify the most frequently occurring terms. A few such phrases have been depicted in Fig.1 which were used to query tweets related to blood donation after removing the stop words. Apart from them, tweets mined by using general medical terms were incorporated to form the complete dataset.

2. **Blood Donation Community(BDC) Dataset** (1889 EBDR, 3268 non-EBDR): The list of users present in the tweets in PDR dataset was obtained. Specific Twitter handles of community blood donation groups were identified from these users and historical tweets from their timeline were extracted for the positive class. For the negative class, past tweets from extraneous users were collected.

3. **Dataset HO:** (741 EBDR, 1072 non-EBDR): This represents the held-out dataset, with tweets collected using both approaches stated above and no overlap with PDR and BDC.

We utilized the traditional query based tweet mining practice which made the PDR dataset more generic in nature. In addition, we employed a Twitter handle supervision technique in BDC dataset to focus on a larger tweet corpus heavily specific to the positive class. Lastly, the held out dataset was created using both the techniques in conjunction for a fair assessment of real world cases. The collected corpus of tweets was filtered down to remove tweets involving non-English text using Ling-Pipe, non-Unicode characters, duplicate tweets, and tweets containing only URL's, images, videos or having less than 3 words.

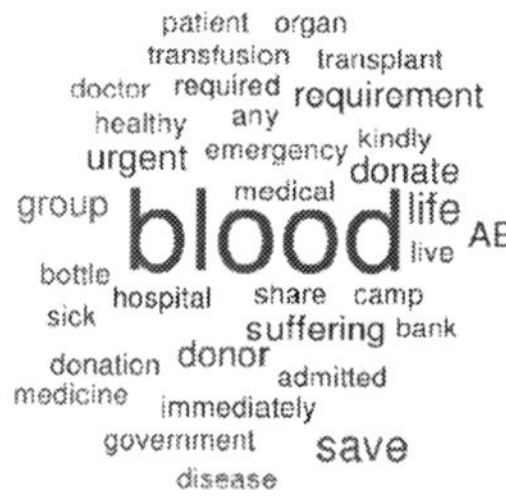

Figure 1: Word cloud of EBDR tweets

3.2 Data Annotation

The datasets were annotated by two independent human annotators. In the case of conflict amongst the annotators, an NLP expert finally assigned the ground truth annotation for ambiguous tweets. A satisfactory agreement between the annotators was inferred from Cohen's Kappa score of 0.86 and the

---

[2]http://www.redcross.org/
[3]https://www.donateblood.com.au/
[4]https://www.blood.co.uk/

Feature Category	Attributes
Linguistic features (L)	Unigram & Bigram presence and count, TF-IDF vector
User metadata (U)	Retweet count, presence of source of posting, presence of place of posting, user friends count, user followers count, user favorites count, user status count
Textual metadata (T)	Count of URL's, hashtags, user mentions and special symbols
Handcrafted features (H)	Presence of name of reference contact, name of place of requirement, contact number, name of hospital/blood bank, blood group required, quantity of blood required, patient disease information

Table 1: Features of tweets in the EBDR dataset.

inter-annotator agreement for the complete EBDR dataset was 89.20%. The classification was done on account of observations stated below:

1. **EBDR Tweets** :

   - Tweets describing a personal critical medical situation indicative of urgent blood requirement within a stipulated timeframe; For instance, *"@BDonors: MOST URGENT B+ve Blood Donors urgently required for a serious cancer patient..."*

   - Tweets appealing for blood donation due to a major crisis or mishap involving casualties and loss of life; e.g., *"Twin blasts in city leave hundreds injured. Request nrby residents having any of A+, B- or O type blood to save precious lives. Contact @username at 99123..."*

2. **Non-EBDR Tweets**:

   - Tweets with no motive to discuss blood donation; e.g., *"We will have to urgently counter this bloody war to sustain our basic living requirements"*.

   - Tweets related to general medical terminology or about general awareness; e.g., *"Iron helps in blood clotting..."*.

   - Promotional content highlighting the usefulness of blood donation to reach out to a target audience. e.g., *"Lets pledge to donate blood every 2 months to help people fighting Leukemia"*.

   - Tweets such as *"We thank @user for registering as #AB- blood donor..."* that portray gratitude for blood donation.

   - Tweets publicizing an offer to donate blood might give rise to contextual bias; e.g., *"I will donate my rare O- blood..."*

Handcrafted Features	PDR	BDC	HO
Name of Reference contact	1117	1513	171
Place of requirement	1188	1844	522
Contact number	1142	1783	541
Hospital/Blood bank	1059	1832	525
Required blood group	1227	1829	701
Patient Disease Description	80	267	109
Quantity of blood required	64	842	156
Total EBDR tweets	**1311**	**1889**	**741**

Table 2: Distribution of hand crafted features of EBDR tweets across Personal Donation Request, Blood Donation Community and Held-Out datasets

4 Feature Modeling

Table 1 presents the complete set of features corresponding to each annotated tweet. The feature set is composed of four constituents: (i) linguistic features, (ii) user metadata, (iii) textual metadata and (iv) handcrafted features. Linguistic features consist of standard unigrams and bigrams as n-grams features along with TF-IDF frequencies that capture the syntactic as well as semantic information. Tweet virality (Cha et al., 2010) and user's network worth (Recuero et al., 2011), measured by the count of friends, followers, favorites and status effect, are necessary parameters to gauge the ability to broadcast emergency messages through the social media network. A common observation during the tweet mining has been the presence of hashtags, URL's and user mentions related to blood donation in EBDR tweets. For instance, hashtags similar to *#SaveLife, #BloodMatters, #HelpEmergency* were prominently present in the positive category of EBDR dataset. Lastly, several handcrafted elements including presence of blood group, blood quantity required, the name of the hospital or blood bank soliciting blood donation on behalf of a patient, disease for which blood transfusion is desired, name, place and contact number of the patient; were extracted by human annotators. Personal details such as user mentions, name, address and phone numbers of patients and tweet posters were anonymized due to privacy concerns of individuals. This resulted in the accumulation of blood donation specific traits as depicted in Table 2.

5 Evaluation

Table 3 shows the performance of datasets BDC, PDR and HO, where the three datasets have been trained using SVM classifier (Chang and Lin, 2011) by taking a combination of one or more feature sets mentioned in Section 4. The train-test split in each case was fixed to 70:30 and stratified five-fold cross-validation performance is reported

Dataset	PDR				BDC				HO			
Feature Set	Accuracy (%)	F1-score	Precision	Recall	Accuracy (%)	F1-score	Precision	Recall	Accuracy (%)	F1-score	Precision	Recall
L	96.22	0.974	**0.979**	0.968	**97.01**	0.958	**0.986**	0.945	97.12	0.974	0.973	**0.986**
U	81.88	0.775	0.759	0.817	51.67	0.564	0.512	0.598	70.18	0.699	0.698	0.701
T	86.62	0.853	0.801	0.887	81.62	0.814	0.812	0.816	85.58	0.855	0.861	0.858
H	96.19	0.975	0.971	0.982	96.59	0.981	0.985	**0.979**	97.01	0.970	**0.983**	0.920
L+H	**96.91**	**0.983**	0.921	**0.986**	96.99	**0.983**	0.985	**0.979**	**97.89**	**0.980**	0.971	0.982
U+T	64.92	0.691	0.780	0.649	77.48	0.732	0.744	0.774	48.48	0.431	0.647	0.484
U+H	87.22	0.879	0.814	0.873	80.80	0.885	0.885	0.896	78.59	0.786	0.853	0.785
T+H	89.49	0.879	0.801	0.836	88.26	0.879	0.823	0.884	89.99	0.875	0.830	0.870
All	75.67	0.759	0.761	0.723	77.11	0.683	0.824	0.771	76.96	0.770	0.840	0.769

Table 3: Results of SVM classifier on PDR, BDC and HO datasets

to account for any imbalance of tweet classes in the datasets that may occur.In each case, linguistic features achieve a marginally better accuracy as compared to the handcrafted features when trained separately, but outperform all other combinations of features when utilized in pair. The extensive under-performance due to inclusion of textual and user metadata prove that these feature sets poorly correlate with the positive class. Dataset BDC consists of a more number of samples having a direct correlation with emergency blood donation requests, as opposed to dataset PDR having a greater abundance of samples relevant to the topic of blood donation. This leads to a higher precision but lower recall in evaluation of Linguistic and Handcrafted feature based PDR dataset. In contrast, the datasets PDR and HO, show a better score, implying the ability to effectively identify posts of EBDR class, thereby reducing the false positive cases. Also, despite the downside of PDR dataset in terms of accuracy, the evaluation metrics follow a similar trend. The best performance in terms of F1-score is shown by using linguistic and handcrafted features in all the three datasets. The HO dataset performs better in terms of accuracy (97.89%) as compared to both PDR and BDC, implying that training classifiers with tweets covering various other topics and aspects increases its robustness towards noise.

5.1 Error Analysis

Some categories of errors that were noticed are:

1. **Rants due to non-availability of blood donors**: Tweets like *"Can't believe we live in a pathetic world, no one came forward to donate a single bottle of B+ve blood ...#HumanityIsDead"* are an example of reactionary posts. Such posts do not belong to EBDR. However the supervised classifiers classify such tweets into the same, making it difficult to separate false requests from genuine cases.

2. **Acknowledgment of blood donation**: The tweet *"We thank @user for registering as #AB- blood donor ..."* was correctly identified by the human annotators but misclassified by the automated classifiers. This can be attributed to the inefficiency of the classifiers to derive contextual meaning from the tweets.

6 Conclusion and Future Work

In this paper, we introduced a robust feature based classification system in addition to an annotated corpus to accurately identify Emergency Blood Donation Request (EBDR) tweets and separate them from other unrelated blood donation communication, referred to as non-EBDR tweets. Given the diverse nature of emergency request tweets, we adopted a two-way corpus construction strategy. We mine three datasets to probe various aspects such as robustness and accuracy and manually annotated them to validate the performance of the proposed classification system. In addition we also perform an analysis of the efficiency of four independent feature sets extracted from the tweets. The results point out that the linguistic features like n-grams and TF-IDF statistics along with handcrafted features related to blood donation requirement are best suited for classification. The EBDR data corpus can benefit researchers in various aspects including but not limited to (i) automatic evaluation of emergency blood donation requests from health posts, (ii) named entity extraction of patient details, blood group and quantity requirement statistics with the help of handcrafted features provided with the tweets, (iii) crisis assessment and management through social media monitoring of medical emergency events and (iv) feature modelling using genetic algorithms as done by Sawhney et al. (2018b,c).

References

David Andre Broniatowski, Mark Dredze, Michael J Paul, and Andrea Dugas. 2015. Using social media to perform local influenza surveillance in an inner-city hospital: a retrospective observational study. *JMIR public health and surveillance*, 1(1).

Meeyoung Cha, Hamed Haddadi, Fabricio Benevenuto, P Krishna Gummadi, et al. 2010. Measuring user influence in twitter: The million follower fallacy. *Icwsm*, 10(10-17):30.

Chih-Chung Chang and Chih-Jen Lin. 2011. Libsvm: a library for support vector machines. *ACM transactions on intelligent systems and technology (TIST)*, 2(3):27.

Carl Lee Hanson, Ben Cannon, Scott Burton, and Christophe Giraud-Carrier. 2013. An exploration of social circles and prescription drug abuse through twitter. *Journal of medical Internet research*, 15(9).

Paul Fontelo MA and MD Eldredge. Development of an adverse drug reaction corpus from consumer health posts.

Debanjan Mahata, Jasper Friedrichs, Rajiv Ratn Shah, and Jing Jiang. 2018a. Did you take the pill?-detecting personal intake of medicine from twitter. *arXiv preprint arXiv:1808.02082*.

Debanjan Mahata, Jasper Friedrichs, Rajiv Ratn Shah, et al. 2018b. # phramacovigilance-exploring deep learning techniques for identifying mentions of medication intake from twitter. *arXiv preprint arXiv:1805.06375*.

Puneet Mathur, Ramit Sawhney, Meghna Ayyar, and Rajiv Shah. 2018a. Did you offend me? classification of offensive tweets in hinglish language. In *Proceedings of the Second Workshop on Abusive Language Online*.

Puneet Mathur, Rajiv Shah, Ramit Sawhney, and Debanjan Mahata. 2018b. Detecting offensive tweets in hindi-english code-switched language. In *Proceedings of the Sixth International Workshop on Natural Language Processing for Social Media*, pages 18–26.

Juan Ramos et al. 2003. Using tf-idf to determine word relevance in document queries. In *Proceedings of the first instructional conference on machine learning*, volume 242, pages 133–142.

Raquel Recuero, Ricardo Araujo, and Gabriela Zago. 2011. How does social capital affect retweets? In *ICWSM*.

Ramit Sawhney, Prachi Manchanda, Raj Singh, and Swati Aggarwal. 2018a. A computational approach to feature extraction for identification of suicidal ideation in tweets. In *Proceedings of ACL 2018, Student Research Workshop*, pages 91–98.

Ramit Sawhney, Puneet Mathur, and Ravi Shankar. 2018b. A firefly algorithm based wrapper-penalty feature selection method for cancer diagnosis. In *International Conference on Computational Science and Its Applications*, pages 438–449. Springer.

Ramit Sawhney, Ravi Shankar, and Roopal Jain. 2018c. A comparative study of transfer functions in binary evolutionary algorithms for single objective optimization. In *International Symposium on Distributed Computing and Artificial Intelligence*, pages 27–35. Springer.

Dealing with medication non-adherence expressions in Twitter

Takeshi Onishi[†], Davy Weissenbacher[‡], Ari Klein[‡], Karen O'Connor[‡], Graciela Gonzalez[‡]
[†]Toyota Technological Institute at Chicago, 6045 South Kenwood, Chicago, IL
[‡]University of Pennsylvania, Philadelphia, PA
[†]tonishi@ttic.edu
[‡]{dweissen, ariklein, karoc, gragon}@pennmedicine.upenn.edu

Abstract

Through a semi-automatic analysis of tweets, we show that Twitter users not only express Medication Non-Adherence (MNA) in social media but also their reasons for not complying; further research is necessary to fully extract automatically and analyze this information, in order to facilitate the use of this data in epidemiological studies.

1 Introduction

Past studies (Claxton et al., 2001) have shown that 50% of medications are not taken as prescribed by patients. This Medication Non-Adherence (MNA) increases morbidity and mortality with an estimated cost of $100-239$ billion per annum to the US healthcare system. The patients' reasons to not comply with treatments are of diverse nature, such as high price for a drug or its negative adverse effect, not trusting the medication, or because they feel better or forget. Healthcare providers have to understand such reasons in order to influence patients' behavior. A major challenge, however, is the difficulty in identifying such reasons. Traditional methods, such as mining clinical records or using pharmacy claims data, or interacting directly with patients through surveys and intervention trials have been found limited to identify MNA reasons (Xie et al., 2017).

With the large adoption during the last decades of Social Media (SM) and the proneness of the SM users to discuss medical habits and share health issues, SM is increasingly regarded as an important source that can provide unique insights into Medication Non-Adherence reasons. For our study, we have chosen Twitter due to the large volume of easily-accessible data.

In this work, through a semi-automatic analysis of 4 million tweets, we show that not only Twitter users clearly express MNA in the social media but also their reasons for not complying with their treatments, which calls for further research to fully automatize our process.

2 Methods

Tweets mention medications in various contexts such as advertising/selling drugs or personal drug experiences. Typically, accounts owned by a company or organization advertise and sell drugs, and individual persons post personal drug experiences such as their prescriptions, their reaction to the drugs, and sometimes their non-adherence.

To determine if Twitter users are mentioning their non adherence to their treatment and their reasons, we manually analyzed an existing corpus of four millions tweets, the Pregnancy corpus. This corpus is composed of ~112,500 timelines[1] of women posting during their pregnancy and collected for the needs of a previous epidemiologic study (Golder et al., 2018).

Two independent methods detected tweets mentioning a MNA. The methods rely on different features related to MNA and were applied in parallel.

Drug names matching: We compiled a list of 103 distinct names of drugs related to HIV and diabetes from Drug.com[2] and eMEDTV[3]. These lists include generic and brand names. We filtered out all tweets which did not contain any drug name from the drug list. We, then, removed all tweets containing a hyperlink, retweets, reply tweets and tweets not written in English. These heuristic rules were inspired by Adrover et al. (2015) and were based on the observation that a majority of tweets containing drug names and a hyperlink were posted by companies commenting web articles, whereas tweets posted by individuals

---

[1]We call a *timeline* the exhaustive set of tweets posted by a user during a given period.

[2]https://www.drugs.com

[3]http://cholesterol.emedtv.com

Proceedings of the 3rd Social Media Mining for Health Applications (SMM4H) Workshop & Shared Task, pages 32–33
Brussels, Belgium, October 31, 2018. ©2018 Association for Computational Linguistics

	Drug	**Pattern**
Tweets matched	377	27
Tweets mentioning MNA	9	9
Reason in the tweet	6	8
Reason in the tweet vicinity	0	1

Table 1: Tweets mentioning a MNA and its reason manually discovered in tweets retrieved by Drug names and Patterns matching methods.

describing their experiences about drugs did not contain hyperlinks.

Patterns matching: We encoded our patterns in REs and searched for all tweets in the corpus. For this preliminary study, we searched for two patterns: all tweets which contain both phrases "stopped taking" and "made me", regardless of the order. The previous heuristic rules, used to remove tweets posted by bots or companies, were not applied on the tweets retrieved by the patterns since, due to the semantic of the patterns, the tweets they retrieved were personal tweets.

Two annotators independently investigated the tweets obtained by both methods. Each annotator judged if the tweets were mentioning a MNA or not for precision. The recall was not estimated because MNA tweets are rare and estimating such frequency even from random samples is practically impossible. For the tweet mentioning an MNA, they looked for the reasons in the users' timelines up to ten days before and after the MNA tweet. A third annotator resolved the disagreements.

3   Results

Table (1) details our results. Despite the limited number of drug names and the small size of our corpus, the drug names matching method retrieved 377 tweets including nine tweets mentioning an MNA. Six of the nine tweets were also describing the reason of the MNA in the tweet. The patterns matching method retrieved 27 tweets including nine MNA tweets. The 27 tweets are exclusive to the 377 tweets retrieved by the first method. The precision of the pattern matching is 9/27 which appears to be more precise compared to that of the drug name matching (9/377). Of these nine tweets retrieved by the pattern matching, one specifies a medication, two specifies a type of medication (*e.g,* pain medication), and the other six use a generalization (*e.g,* pills) or a pronoun to refer to a medication mentioned elsewhere in the users' timelines. Due to the patterns searched, all of the tweets also mention the reasons, except for one tweet that is truncated and mentions the reason in the subsequent post. The other 18 tweets retrieved by the patterns either did not refer to a type of medication (*e.g,* birth control, prenatal vitamins) or used generalizations or pronouns to refer to medications for which we did not discover the referent.

4   Conclusion

Two semi-automatic processes successfully isolated 18 tweets in total from four millions of tweets where Twitter users explicitly report their MNA. Additionally, we found that users are also more likely to explain their failure to comply in the same MNA tweets or in the following tweets. These results showed the potential of Twitter for understanding patients' behavior at a large scale and justify further research to extract and analyze automatically the MNA reasons. To increase the number of tweets retrieved, we will listen in real-time tweets from the stream of Twitter, searching for all drugs names using a Drug Name Recognizer and a manually expended set of patterns.

References

Cosme Adrover, Todd Bodnar, Zhuojie Huang, Amalio Telenti, and Marcel Salathé. 2015. Identifying Adverse Effects of HIV Drug Treatment and Associated Sentiments Using Twitter. *JMIR Public Health and Surveillance*, 1(2).

Ami J. Claxton, Joyce Cramer, and Courtney Pierce. 2001. A Systematic Review of the Associations Between Dose Regimens and Medication Compliance. *CLINICAL THERAPEUTICS*, 23(8):1296–1310.

Su Golder, Stephanie Chiuve, Davy Weissenbacher, Ari Klein, Karen O'Connor, Martin Bland, Murray Malin, Mondira Bhattacharya, Linda Scarazzini, and Graciela Gonzalez-Hernandez. 2018. Pharmacoepidemiologic evaluation of birth defects from health-related postings in social media during pregnancy. *Drug Safety [Submitted]*.

Jiaheng Xie, Xiao Liu, Daniel Dajun Zeng, and Xiao Fang. 2017. Understanding medication nonadherence from social media: A sentiment-enriched deep learning approach. *SSRN Electronic Journal*.

Detecting Tweets Mentioning Drug Name and Adverse Drug Reaction with Hierarchical Tweet Representation and Multi-Head Self-Attention

Chuhan Wu[1], Fangzhao Wu[2], Junxin Liu[1], Sixing Wu[1], Yongfeng Huang[1] and Xing Xie[2]

[1] Department of Electronic Engineering, Tsinghua University, Beijing 100084, China
[2]Microsoft Research Asia
{wuch15,ljx16,wu-sx15,yfhuang}@mails.tsinghua.edu.cn
fangzwu,xing.xie@microsoft.com

Abstract

This paper describes our system for the first and third shared tasks of the third Social Media Mining for Health Applications (SMM4H) workshop, which aims to detect the tweets mentioning drug names and adverse drug reactions. In our system we propose a neural approach with hierarchical tweet representation and multi-head self-attention (HTR-MSA) for both tasks. Our system achieved the first place in both the first and third shared tasks of SMM4H with an F-score of 91.83% and 52.20% respectively.

1 Introduction

Social media services such as Twitter have become important platforms for information sharing and dissemination. Automatically detecting tweets which mentions drug names (DNs) and adverse drug reactions (ADRs) at a large scale is an interesting research topic and has many important applications such as pharmacovigilance (Sarker and Gonzalez, 2015; Han et al., 2017; Weissenbacher et al., 2018). However, tweets are very noisy and informal, and full of misspellings (e.g., "aspirn" for "aspirin") and user-created abbreviations (e.g., "COC" for "Cocaine"). In addition, many DN and ADR mentions are context-dependent. For example, "I take Vitamin C after meals" is a tweet mentioning drug name, but the tweet "Vitamin C is good for health" is not. Thus, the detection of DN and ADR mentioning tweets is very challenging.

In order to facilitate the research on automatic detection of tweets mentioning DN and ADR, two related shared tasks were released by the third Social Media Mining for Health Applications (SMM4H) workshop[1] (Weissenbacher et al., 2018). Task 1 aims to classify whether a tweet mentions any drug names or dietary supplement.

Task 3 aims to classify whether a tweet contains adverse drug reaction mention. We designed a neural approach with hierarchical tweet representation and multi-head self-attention (HTR-MSA) to participate in these two tasks. Our hierarchical tweet representation model first learns word representations from characters using convolutional neural network (CNN) and then learns tweet representations from words using a combination of Bi-directional long-short term memory (Bi-LSTM) network and CNN. In addition, we incorporated additional features to enhance the word representations, including pre-trained word embedding, part-of-speech (POS) tag embedding, sentiment features based on sentiment lexicons and lexicon features extracted from medical lexicons. Besides, we applied multi-head self-attention mechanism to our approach to enhance the contextual representations of words by capturing the interactions between all words in tweets. Our system achieved 91.83% F-score in Task 1 and 52.20% F-score in Task 3, and ranked 1st in both task. The codes of our system are publicly available[2].

2 Our Approach

The architecture of our HTR-MSA model is shown in Fig. 1. It contains three major modules, i.e., word representation, tweet representation and tweet classification.

2.1 Word Representation

In order to handle the massive misspellings and user-created abbreviations of drug names in tweets, we propose to learn word representations from characters. There are three sub-modules in the word representation module.

The first one is character embedding, which converts each word from a sequence of characters into a sequence of low-dimensional dense vectors

---

[1]https://healthlanguageprocessing.org/smm4h/

[2]https://github.com/wuch15/SMM4H_THU_NGN

Proceedings of the 3rd Social Media Mining for Health Applications (SMM4H) Workshop & Shared Task, pages 34–37
Brussels, Belgium, October 31, 2018. ©2018 Association for Computational Linguistics

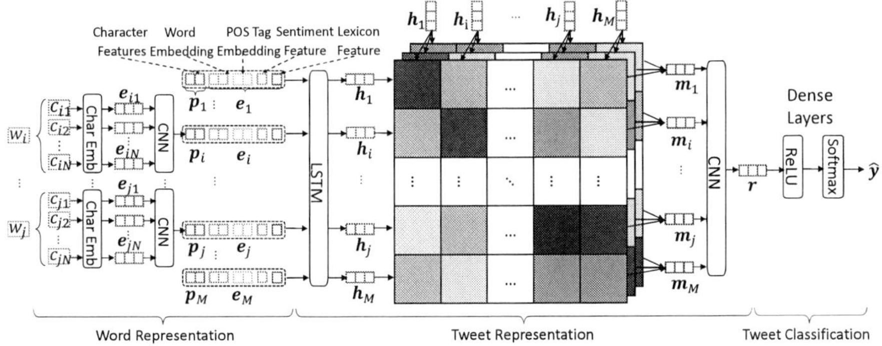

Figure 1: Architecture of our HTR-MSA model.

using a character embedding matrix. The second one is a character-level CNN network to learn contextual representations of characters. CNN is effective to capture local context information. Since many drug names contain specific character combinations (e.g., "benz" and "acid"), we apply CNN to learn contextual character representations by capturing the local information of neighbor characters. We use max-pooling operation to the feature maps generated by multiple filters in CNN to select the most significant features to build word representations based on characters.

The third one is feature concatenation, where the word representation learned from characters is concatenated with additional word features to build the final word representation vector. The first additional feature is word embeddings which are pre-trained on a large corpus and contain rich semantic information of words. According to previous studies (Sarker and Gonzalez, 2015), sentiment information and medical lexicons are very important for DN and ADR detection. Therefore, we incorporate words' sentiment scores extracted from SentiWordNet 3.0 sentiment lexicon [3] and their appearance in the SIDER 4.1 medical lexicon[4] into their representation vectors. In addition, since DN and ADR mentions usually have specific POS tags (e.g., nouns), we also incorporate the embeddings of their POS tags . The final representation vector of a word is a concatenation of its character-based representation, word embedding, POS tag embedding, sentiment scores and lexicon appearance.

2.2 Tweet Representation

The tweet representation module aims to learn the representation vectors of tweets from their words.

It also contains three sub-modules.

The first one is a Bi-LSTM network (Graves and Schmidhuber, 2005). Long-distance information is very important for the detection of tweets mentioning DN and ADR. For example, the tweet "I took amoxicillin last night, but I find I'm so tired today" contains an ADR mention "tired", which has a long distance to the drug name "amoxicillin". LSTM is an effective network to capture long-distance information. We use Bi-LSTM network in our approach. It can capture the context information from both directions and output the hidden states at each position. Denote the hidden states of words in a tweet as $\mathbf{H} = [\mathbf{h}_1, ..., \mathbf{h}_M]$, where M is sentence length.

The second sub-module is multi-head self-attention network. In most of existing attention mechanisms the attention weight of a word is computed only based on its hidden representation, and the relationships between different words in a text cannot be modeled. Usually, many DN and ADR mentions are context-dependent and the interactions between words are very important to detect the DN and ADR mentioning tweets. Self-attention is an effective way to capture the useful interactions between words in texts (Vaswani et al., 2017). In addition, a word may interact with multiple words. For example, in the tweet "I forgot to take aspirin and I'm in huge pain", the interaction of "aspirin" with "forgot" and the interaction of "aspirin" with"pain" are both important for ADR mention detection. Thus, we propose to use multi-head self-attention mechanism (Vaswani et al., 2017) to learn better hidden representations of words by modelling their interactions with multiple words. In this layer, the representation vector $\mathbf{m}_{i,j}$ of the j_{th} word learned by the i_{th} attention head is computed by a weighted summation of $\mathbf{H}$ as follows:

[3] http://sentiwordnet.isti.cnr.it/ (last access: Jul 19.)
[4] http://sideeffects.embl.de/ (last access: Jul 20.)

$$\hat{\alpha}_{j,k}^i = \mathbf{h}_j^T \mathbf{U}_i \mathbf{h}_k, \tag{1}$$

$$\alpha_{j,k}^i = \frac{\exp(\hat{\alpha}_{j,k}^i)}{\sum_{m=1}^{M} \hat{\alpha}_{j,m}^i}, \tag{2}$$

$$\mathbf{m}_{i,j} = \mathbf{W}_i(\sum_{m=1}^{M} \alpha_{j,m}^i \mathbf{h}_m), \tag{3}$$

where $\mathbf{U}_i$ and $\mathbf{W}_i$ are the parameters of the i_{th} self-attention head, and $\alpha_{j,k}^i$ represents the relative importance of the interaction between the j_{th} and k_{th} words. In this way, the representation of each word is learned by utilizing the hidden representations of all words in the same text and modeling the interactions between this word with all other words. The multi-head representation $\mathbf{m}_j$ of the j_{th} word is the concatenation of the outputs from h different self-attention heads, i.e., $\mathbf{m}_j = [\mathbf{m}_{1,j}; \mathbf{m}_{2,j}; ...; \mathbf{m}_{h,j}]$.

The third sub-module is a word-level CNN network with max-pooling operation. Since many drug names contain specific word combinations (e.g., salicylic acid and acetic acid), local contextual information between words is important for DN and ADR detection. We apply CNN to the sequence of hidden representations of words in each tweet, and the final representation vector of a tweet $\mathbf{r}$ is obtained from the results of max-pooling on the CNN feature maps.

2.3 Tweet Classification

The tweet classification module is used to classify whether a tweet mentions DN or ADR. It contains two dense layers with ReLU and softmax activation functions respectively. The predicted label $\hat{y}$ of a tweet is computed as:

$$\mathbf{r}' = ReLU(\mathbf{U}_1 \mathbf{r} + \mathbf{b}_1), \tag{4}$$

$$\hat{\mathbf{y}} = softmax(\mathbf{U}_2 \mathbf{r}' + \mathbf{b}_2), \tag{5}$$

where $\mathbf{U}_1$, $\mathbf{U}_2$, $\mathbf{b}_1$, $\mathbf{b}_2$ are the parameters for DN and ADR mention classification. The loss function $\mathcal{L}$ used for model training is crossentropy:

$$\mathcal{L} = -\sum_{i=1}^{N} \sum_{k=1}^{K} y_k \log(\hat{y}_k), \tag{6}$$

where $y_{i,k}$ and $\hat{y}_{i,k}$ are gold label and predicted label for the i_{th} tweet in the k_{th} label category. N is the number of labeled tweets.

3 Experiments

3.1 Datasets and Experimental Settings

The datasets provided by Task 1 and Task 3 in the shared tasks of the third SMM4H workshop (Weissenbacher et al., 2018) were used in our experiments. The first one is for detection of tweets mentioning DNs (denoted as DN). It contains 9,622 tweet IDs (4,975 positive and 4,647 negative samples) for training, and 5,382 tweet for test. The third one is for detection of tweets mentioning ADRs (denoted as ADR). It contains 25,598 tweet IDs (2,223 positive and 23,375 negative samples) for training, and 5,000 tweets for test. Since many tweets are not available now, we only crawled 9,065 and 16,694 tweets for training in DN and ADR respectively using these IDs.

In our experiments, we use the 400-dim pre-trained word embeddings released by Godin et al. (2015). The Bi-LSTM network has 2×200 units. The CNN network has 400 filters with window size of 3. There are 16 heads in the multi-head self-attention network, and the output dimension of each head is 16. RMSProp is selected as the optimizer. Since the negative samples are dominant in the ADR dataset, we use the over-sampling strategy (Weiss et al., 2007) to balance the number of positive and negative samples. Besides, in order to further improve the performance of our approach, we incorporate the ensemble strategy by independently training our model for 10 times and using the average prediction results. The performance metric is F-score on positive samples.

3.2 Performance Evaluation

In this section, we evaluate the performance of our approach by comparing it with baseline methods, including: (1) SVM, support vector machine with word unigram features (Sarker and Gonzalez, 2015); (2) CNN, convolutional neural network (Huynh et al., 2016); (3) LSTM, Bi-LSTM network (Huynh et al., 2016); (4) CRNN, combining CNN and LSTM (Huynh et al., 2016); (5) RCNN, combining LSTM and CNN (Huynh et al., 2016); (6) HTR, our basic hierarchical tweet representation model without self-attention; (7) HTR-MSA, our hierarchical tweet representation model with multi-head self-attention; (8) HTR-MSA-ens, using an ensemble of our HTR-MSA models. For fair comparisons, we use the same additional word features with our approach in all baseline methods. We conducted 10-fold cross-validation on the labeled tweets and the results are summarized in Table 1. According to Table 1, our approach can outperform all the baseline methods. This may be because in our approach we learn word representations from not only the word embeddings but also the characters in words.

Method	DN	ADR
SVM	88.20	47.20
CNN	89.16	48.56
LSTM	88.78	48.28
CRNN	89.10	48.44
RCNN	89.31	48.75
HTR	89.80	49.49
HTR-MSA	90.57	50.55
HTR-MSA-ens	**91.85**	**52.48**
HTR-MSA-ens*	**91.83**	**52.20**

Table 1: The performance of different methods in the DN and ADR detection task. *Results on the test set.

Thus, our approach can be more robust to the massive misspellings of drug names in tweets and can mitigate the influence of out-of-vocabulary words. In addition, by comparing the results of HTR-MSA and HTR, we find that the multi-head self-attention network is helpful to improve the performance of our approach. This may be because the global context information is very important for detecting tweets mentioning DNs and ADRs and the multi-head self-attention network can effectively capture the interactions between words within a tweet. Besides, ensemble strategy can further improve the performance of our approach. It indicates that a more robust system can be built for detecting tweets mentioning drug names and adverse drug reactions using the ensemble of multiple models independently trained using our approach.

3.3 Influence of Additional Word Features

In this section, we conducted experiments to explore the effectiveness of additional word features and the results are shown in Table 2. According to Table 2, each kind of additional word feature, such as word embedding, POS tag embedding, sentiment score and medial lexicon features, is effective to improve the performance of our approach. In addition, among these additional word features word embedding seems to be most useful. This is probably because that pre-trained word embeddings can provide rich semantic information of words, which is important for detecting tweets mentioning DNs and ADRs.

Feature	DN	ADR
All	90.57	50.55
-Word embedding	86.45	46.29
-POS tag embedding	90.26	50.31
-Sentiment scores	90.33	50.29
-Lexicon feature	89.94	50.10

Table 2: Effectiveness of additional word features.

4 Conclusion

In this paper, we introduce our system participating in the first and the third shared tasks in the 3rd SMM4H workshop. We propose a neural approach with hierarchical tweet representation and multi-head self-attention to detect tweets mentioning DNs and ADRs. Our system achieved the first place in both tasks.

Acknowledgments

This work was supported in part by the National Key Research and Development Program of China under Grant 2016YFB0800402 and the National Natural Science Foundation of China under Grant U1705261, U1536207, U1536201 and U1636113.

References

Fréderic Godin, Baptist Vandersmissen, Wesley De Neve, and Rik Van de Walle. 2015. Multimedia lab @ acl wnut ner shared task: Named entity recognition for twitter microposts using distributed word representations. In *WNUT*, pages 146–153.

Alex Graves and Jürgen Schmidhuber. 2005. Framewise phoneme classification with bidirectional lstm and other neural network architectures. *Neural Networks*, 18(5-6):602–610.

Sifei Han, Tung Tran, Anthony Rios, and Ramakanth Kavuluru. 2017. Team uknlp: Detecting adrs, classifying medication intake messages, and normalizing adr mentions on twitter. In *SMM4H@ AMIA*, pages 49–53.

Trung Huynh, Yulan He, Alistair Willis, and Stefan Rueger. 2016. Adverse drug reaction classification with deep neural networks. In *COLING Technical Papers*, pages 877–887.

Abeed Sarker and Graciela Gonzalez. 2015. Portable automatic text classification for adverse drug reaction detection via multi-corpus training. *Journal of biomedical informatics*, 53:196–207.

Ashish Vaswani, Noam Shazeer, Niki Parmar, Jakob Uszkoreit, Llion Jones, Aidan N Gomez, Łukasz Kaiser, and Illia Polosukhin. 2017. Attention is all you need. In *NIPS*, pages 5998–6008.

Gary M Weiss, Kate McCarthy, and Bibi Zabar. 2007. Cost-sensitive learning vs. sampling: Which is best for handling unbalanced classes with unequal error costs? *DMIN*, 7:35–41.

Davy Weissenbacher, Abeed Sarker, Michael Paul, and Graciela Gonzalez-Hernandez. 2018. Overview of the third social media mining for health (smm4h) shared tasks at emnlp 2018. In *EMNLP*.

Classification of medication-related tweets using stacked bidirectional LSTMs with context-aware attention

Orest Xherija

Department of Linguistics
University of Chicago
Chicago, IL, USA
`orest.xherija@uchicago.edu`

Abstract

This paper describes the system that team `UChicagoCompLx` developed for the 2018 Social Media Mining for Health Applications (SMM4H) Shared Task. We use a variant of the Message-level Sentiment Analysis (MSA) model of (Baziotis et al., 2017), a word-level stacked bidirectional Long Short-Term Memory (LSTM) network equipped with attention, to classify medication-related tweets in the four subtasks of the SMM4H Shared Task. Without any subtask-specific tuning, the model is able to achieve competitive results across all subtasks. We make the datasets, model weights, and code publicly available[1].

1 Introduction

The Shared Task of the 2018 Social Media Mining for Health Applications (SMM4H) workshop (Weissenbacher et al., 2018) proposed four subtasks in the domain of social media mining for health monitoring and surveillance. From a Natural Language Processing (NLP) viewpoint, these tasks present a considerable challenge since the nature of social media posts requires dealing with both a significant level of language variation and a widespread presence of noise (spelling mistakes, syntactic errors etc). Any classifier designed for this textual domain should take into account the above intricacies and should, furthermore, be able to deal with with semantic complexities in the various ways people express medication-related concepts and outcomes.

To address these challenges, we use a variant of the Message-level Sentiment Analysis (MSA) model of (Baziotis et al., 2017), originally developed for sentiment analysis of Twitter posts, to classify tweets in all four subtasks. The

model is a word-level stacked bidirectional LSTM (BiLSTM) with context-aware attention that uses word-embeddings pretrained by (Baziotis et al., 2017) on a corpus of $\approx$ 330M tweets. Without additional hyperparameter tuning or subtask-specific modifications, the model outperforms the average of all submitted systems in subtasks 1 and 4 and achieves first place (by a F1-score margin of 0.234 from the next team) in subtask 2. In subtask 3 our model was placed 6th out of 9 systems.

In the following sections, we introduce the datasets, discuss preprocessing steps we took, present the model and its training setup, report results, and conclude with potential avenues for future research.

2 Datasets

In this section, we describe the datasets of each subtask. Subtasks 1, 3 and 4 are binary classification problems while subtask 2 is a three-class classification problem. The data was manually annotated by the organizers.

Subtask 1 is about the automatic detection of posts mentioning the name of a drug or dietary supplement, as defined by the United States Food and Drug Administration (FDA). A tweet is assigned label 1 if it contains the name of one or more drugs or supplements and 0 otherwise. **Subtask 2** poses the challenge of automatic classification of posts describing medication intake. A tweet is assigned label 1 if "the user clearly expresses a personal medication intake/consumption", 2 if the tweet suggests (without certainty) that "the user may have taken the medication", and 3 if the tweet mentions medication names but does not indicate personal intake. **Subtask 3** concerns the automatic classification of posts mentioning an adverse drug reaction (ADR). A tweet is assigned label 1 if it men-

---

[1] `https://github.com/orestxherija/smm4h2018`

Proceedings of the 3rd Social Media Mining for Health Applications (SMM4H) Workshop & Shared Task, pages 38–42
Brussels, Belgium, October 31, 2018. ©2018 Association for Computational Linguistics

	1	2	3	4
training	7011	13791	21062	6956
validation	780	1533	2341	CV
evaluation	5382	5000	5000	161

Table 1: Examples per split per task. CV indicates cross-validation, so no validation set was held out.

tions an ADR and 0 otherwise. Finally, **Subtask 4** deals with the automatic detection of posts mentioning vaccination behavior related to influenza vaccines. The annotators were asked the question "Does this message indicate that someone received, or intended to receive, a flu vaccine?" and a tweet was assigned label 1 if the answer was affirmative and 0 otherwise. Subtasks 1, 3 and 4 are evaluated using the F1-score for the positive class while subtask 2 uses the micro-averaged F1-score for classes 1 and 2. Subtask 1 is additionally evaluated on precision and recall for the positive class.

Due to Twitter privacy policies, the training set for any subtask did not contain the actual tweet text. To obtain said text, participants were provided with the tweet ID of each dataset example along with a script to use for downloading the text using this ID. The process inevitably resulted in fewer tweets than the number of IDs contained in the original dataset, primarily because a number of tweets had been removed (either by the users themselves, or by Twitter because e.g. the user deleted his account) while others failed to download (due to e.g. lag issues when requesting the HTML of the tweet). To avoid such issues in the evaluation datasets, the organizers decided to provide the tweet text along with the ID. Table [1] provides a short summary of the number of tweets that were available to our team for each subtask.

3 Pre-processing

We applied identical preprocessing to all datasets. We replaced Twitter specific strings with appropriate tokens (e.g. emojis were replaced by **$EMOJI$**, numbers were replaced by **$NUMBER$**, website urls by **URL** etc) to reduce the vocabulary size and to ameliorate the noisy nature of the text. All non-alphanumeric characters and all tokens that were too short (fewer than 2 characters) or too long (more than 15 characters) were removed. Finally, all text was converted to lower case and any excess whitespace (i.e. newlines and tabs) was removed.

4 Model description

4.1 Model architecture

We use a variant of the Message-level Sentiment Analysis (MSA) model of (Baziotis et al., 2017). The model consists of two stacked BiLSTMs with a context-attention mechanism à la (Yang et al., 2016) that identifies the maximally informative words for each label. We describe subsequently the individual network layers.

The input is a tweet, regarded as a sequence of words, which is projected to a vector space of fixed size via the **Embedding Layer**. The weights of the embedding layer are initialized using pre-trained word embeddings that (Baziotis et al., 2017) trained on a Twitter corpus of approximately $\approx 330M$ tweets. We opt for these embeddings instead of the standard Word2Vec (Mikolov et al., 2013a,b) ones since they have been trained on a similar textual domain to the tasks at hand.

A **LSTM Layer** placed on top of the embedding layer takes as input the embedding weights and produces a representation $\{h_i\}_{i=1}^T$ where, h_i is the hidden state of the LSTM at time-step i, intuitively corresponding to a summary of all the information of the sentence (viewed as a sequence $\{w_i\}_{i=1}^T$ of words) up to w_i. This constitutes a forward LSTM. Since we are using a bidirectional LSTM, we also have an LSTM that scans the sequence of words in the reverse direction. The final representation of a word is produced by concatenating the representations from the forward and backward LSTM:

$$h_i = \overrightarrow{h_i} || \overleftarrow{h_i} \qquad (1)$$

where $||$ denotes the concatenation operator. We opt for a stacked BiLSTM, and consequently we place an additional BiLSTM layer on top of the preceding layer. The motivation for this choice comes from the literature on the interpretation of hidden states of Recurrent Neural Networks (RNNs) (Belinkov et al., 2017; Belinkov, 2018) in which it has been claimed that deeper layers are able to learn more abstract semantic representations of sentences, thus achieving superior performance in downstream tasks.

To account for the fact that not all words contribute equally to the assignment of a label, we place an **Attention Layer** on top of the BiLSTMs following work like (Sutskever et al., 2014) who successfully used attention mechanisms for sequence-to-sequence neural machine translation. We use context-attention, following (Yang et al., 2016). A context vector u_h is initialized and is governed by the following update equations:

$$e_i = \tanh(W_h h_i + b_h) \tag{2}$$

$$a_i = \frac{\exp(e_i^\top u_h)}{\sum_{j=1}^{T} \exp(e_j^\top u_h)} \qquad \sum_{i=1}^{T} a_i = 1 \tag{3}$$

$$r = \sum_{i=1}^{T} a_i h_i \qquad r \in \mathbb{R}^{2L} \tag{4}$$

where W_h, b_h and u_h are learned parameters, h_i is the concatenation of the representations of the forward and backward LSTM, introduced in equation (1), and L is the number of cells in one LSTM layer.

Finally, we feed the representation r produced by the attention layer to a **Dense Layer** with sigmoid activation (softmax for subtask 2) and obtain a probability distribution over the classes. If the probability assigned to a tweet is greater than 0.5 we assign label 1, otherwise we assign 0.

4.2 Training setup

We train the model to minimize the negative log-likelihood loss using back-propagation with stochastic gradient descent and mini-batch size of 50. We use the Adam optimizer (Kingma and Ba, 2015) with gradient norm clipping (Pascanu et al., 2013) at 1. For subtasks 1, 2 and 3 we use a $90-10$ train-validation split, while for subtask 4 we use $10-$fold stratified cross-validation in consideration of the very small test set. Table [1] summarizes the information on train-validation splits.

4.3 Regularization

To make the model more robust to over-fitting, we employ, following (Baziotis et al., 2017), a number of regularization techniques. We add Gaussian noise at the embedding layer and use dropout (Srivastava et al., 2014) to ignore the signal from a set of randomly selected neurons in the network. Dropout is also applied after each LSTM layer as well as to the recurrent connections of the LSTM (Gal and Ghahramani, 2016). L_2 regularization along with class weights are applied to the loss function to prevent overly large weights and to account for class imbalance. Class weights are computed as follows: assuming that $\vec{x}$ is the vector of class counts, the weights are defined

as $w_i = \max(\vec{x})/x_i$ for any class i. Finally, early-stopping (Caruana et al., 2001) is employed to terminate training after the validation loss has stopped decreasing.

4.4 Hyperparameter tuning

We use the similar hyperparameters to (Baziotis et al., 2017). In particular, we use 150 as the size of the LSTM hidden states (300 in total since we are using a BiLSTM), the Gaussian noise parameter is set to $\sigma = 0.3$, dropout rate on top of the embedding layer is set to 0.3 and dropout rate on top of the LSTM layers is set to 0.5. Dropout at the recurrent connections is also set to 0.3. L_2 regularization at the loss function is set to 0.0001. Finally, we initialize the learning rate at 0.001. Departing from (Baziotis et al., 2017), we use word embeddings of dimension 100. Vocabulary size and maximum sequence length are set to 7000 and 50 respectively for all subtasks and the patience level for early-stopping is set to 0.001 in 5 epochs.

5 Experiments and results

5.1 Experimental setup

The model was developed using Keras[2] with the Tensorflow (Abadi et al., 2016) backend. For data preparation and processing we use Scikit-learn (Pedregosa et al., 2011). Given the small size of the datasets, we do not use GPUs for training the model. A standard 8-core CPU is sufficient. Finally, for designing the network architecture, we use part of the code released by (Baziotis et al., 2017)[3].

5.2 Results

For subtasks 1 and 4, the organizers chose to disclose to each team only their respective score along with the average score of all submitted systems. These results are summarized in Table [2]. Our system performed better than the average in both subtasks, considerably so in subtask 1.

For subtasks 2 and 3, the organizers released the complete leaderboards, presented in Tables [3] and [4] respectively. Our system greatly outperformed all other systems by a significant margin in subtask 2. In subtask 3, our system ranked 6th (out of 9 participants), potentially because the other teams developed specialized systems for the particular

---

[2]https://keras.io/
[3]https://github.com/cbaziotis/
datastories-semeval2017-task4

	P	R	F1
Subtask-1	**0.937**	**0.891**	**0.914**
	(0.890)	(0.872)	(0.880)
Subtask-4	0.791	0.923	**0.852**
	(0.826)	(0.858)	(0.840)

Table 2: Results on the evaluation set for subtasks 1 and 4. Average score of all participating systems in parentheses. Metric is F1-score for class 1. For subtask 1, precision and recall for class 1 are also used for evaluation.

	P	R	F1
UChicagoCompLx	**0.654**	**0.783**	**0.713**
Light	0.492	0.467	0.479
Tub-Oslo	0.464	0.466	0.465
IRISA_team	0.434	0.501	0.465
IIT_KGP	0.408	0.407	0.408
UZH	0.371	0.437	0.401
CLaC	0.402	0.366	0.383
Techno	0.327	0.432	0.372

Table 3: Subtask 2 final leaderboard. Metric is micro-averaged F1-score for classes 1 and 2.

subtask while we opted for a general model that can be used without modifications in all four subtasks.

6   Conclusion and future directions

We demonstrated that the variant of the MSA model of (Baziotis et al., 2017) performs competitively when applied to the domain of medication-related short text classification. Without hyperparameter tuning, major architectural modifications, or task-specific adjustments, the model obtained competitive results in subtasks 1 and 4 and ranked

	P	R	F1
THU_NGN	0.442	0.636	**0.522**
IRISA_team	0.378	**0.649**	0.478
UZH	0.455	0.436	0.445
Tub-Oslo	**0.638**	0.317	0.424
Art	0.332	0.547	0.413
UChicagoCompLx	0.370	0.464	0.411
CIC-NLP	0.314	0.529	0.394
Techno	0.434	0.344	0.383
IIT_KGP	0.189	0.643	0.292

Table 4: Subtask 3 final leaderboard. Metric is F1-score for class 1.

first in subtask 3, greatly outperforming all other models in terms of precision, recall and F1-score. The model's performance in this Shared Task is further testament to the ability of attentive RNNs to perform at state-of-the-art level in short text classification where individual word-meaning is essential.

In the future, we aim to investigate whether ensembles of word- and character-level attentive RNNs can perform even better. The benefits of ensembling for text classification can be seen in numerous NLP tasks ranging from Natural Language Inference (Gong et al., 2018, among many others) to product categorization (Skinner, 2018). Word-level models perform well in capturing aspects of the semantics (Belinkov et al., 2017) while character-level models succeed in capturing syntactic information about the text. Ensembles of these diverse types of models can potentially lead to improved performance.

A second avenue to pursue would be multi-task learning, an area of active research that has shown promising results in text classification (Liu et al., 2016, 2017, among others). Given that all subtasks are nearly identical in nature (all but one of them being binary classification problems) and share a highly overlapping lexicon, they provide an excellent ground for testing the merits of multi-task learning.

Acknowledgments

This work was completed in part with resources provided by the University of Chicago Research Computing Center, whose contribution we gratefully acknowledge.

References

Martín Abadi, Paul Barham, Jianmin Chen, Zhifeng Chen, Andy Davis, Jeffrey Dean, Matthieu Devin, Sanjay Ghemawat, Geoffrey Irving, Michael Isard, Manjunath Kudlur, Josh Levenberg, Rajat Monga, Sherry Moore, Derek G. Murray, Benoit Steiner, Paul Tucker, Vijay Vasudevan, Pete Warden, Martin Wicke, Yuan Yu, and Xiaoqiang Zheng. 2016. TensorFlow: A System for Large-Scale Machine Learning. In *12th USENIX Symposium on Operating Systems Design and Implementation (OSDI 16)*, pages 265–283, Savannah, GA, USA. USENIX Association.

Christos Baziotis, Nikos Pelekis, and Christos Doulkeridis. 2017. DataStories at SemEval-2017 Task 4: Deep LSTM with Attention for Message-level and

Topic-based Sentiment Analysis. In *Proceedings of the 11th International Workshop on Semantic Evaluation (SemEval-2017)*, pages 747–754, Vancouver, Canada. Association for Computational Linguistics.

Yonatan Belinkov. 2018. *On Internal Language Representations in Deep Learning: An Analysis of Machine Translation and Speech Recognition*. Ph.D. thesis, Massachusetts Institute of Technology.

Yonatan Belinkov, Lluís Màrquez, Hassan Sajjad, Nadir Durrani, Fahim Dalvi, and James Glass. 2017. Evaluating Layers of Representation in Neural Machine Translation on Part-of-Speech and Semantic Tagging Tasks. In *Proceedings of the Eighth International Joint Conference on Natural Language Processing*, volume 1, pages 1–10, Taipei, Taiwan. Asian Federation of Natural Language Processing.

Rich Caruana, Steve Lawrence, and Lee Giles. 2001. Overfitting in Neural Nets: Backpropagation, Conjugate Gradient, and Early Stopping. In Todd K. Leen, Thomas G. Dietterich, and Volker Tresp, editors, *Advances in Neural Information Processing Systems*, volume 13, pages 402–408. MIT Press, Denver, CO, USA.

Yarin Gal and Zoubin Ghahramani. 2016. Dropout As a Bayesian Approximation: Representing Model Uncertainty in Deep Learning. In *Proceedings of The 33rd International Conference on Machine Learning*, volume 48 of *Proceedings of Machine Learning Research*, pages 1050–1059, New York, NY, USA. PMLR.

Yichen Gong, Heng Luo, and Jian Zhang. 2018. Natural Language Inference over Interaction Space. In *International Conference on Learning Representations*, Vancouver, Canada.

Diederik P. Kingma and Jimmy Ba. 2015. Adam: A Method for Stochastic Optimization. In *International Conference on Learning Representations*, San Diego, CA, USA.

Pengfei Liu, Xipeng Qiu, and Xuanjing Huang. 2016. Deep Multi-Task Learning with Shared Memory for Text Classification. In *Proceedings of the 2016 Conference on Empirical Methods in Natural Language Processing*, pages 118–127, Austin, Texas. Association for Computational Linguistics.

Pengfei Liu, Xipeng Qiu, and Xuanjing Huang. 2017. Adversarial Multi-task Learning for Text Classification. In *Proceedings of the 55th Annual Meeting of the Association for Computational Linguistics*, volume 1, pages 1–10, Vancouver, Canada. Association for Computational Linguistics.

Tomas Mikolov, Kai Chen, Greg Corrado, and Jeffrey Dean. 2013a. Efficient Estimation of Word Representations in Vector Space. In *International Conference on Learning Representations*, Scottsdale, AZ, USA.

Tomas Mikolov, Ilya Sutskever, Kai Chen, Greg Corrado, and Jeffrey Dean. 2013b. Distributed Representations of Words and Phrases and Their Compositionality. In Christopher J. C. Burges, Léon Bottou, Max Welling, Zoubin Ghahramani, and Kilian Q. Weinberger, editors, *Advances in Neural Information Processing Systems*, volume 26, pages 3111–3119. Curran Associates, Inc., Lake Tahoe, CA, USA.

Razvan Pascanu, Tomas Mikolov, and Yoshua Bengio. 2013. On the Difficulty of Training Recurrent Neural Networks. In *Proceedings of the 30th International Conference on Machine Learning*, volume 28 of *Proceedings of Machine Learning Research*, pages 1310–1318, Atlanta, GA, USA. PMLR.

Fabian Pedregosa, Gaël Varoquaux, Alexandre Gramfort, Vincent Michel, Bertrand Thirion, Olivier Grisel, Mathieu Blondel, Peter Prettenhofer, Ron Weiss, Vincent Dubourg, Jake VanderPlas, Alexandre Passos, David Cournapeau, Matthieu Brucher, Matthieu Perrot, and Edouard Duchesnay. 2011. Scikit-learn: Machine Learning in Python. *Journal of Machine Learning Research*, 12:2825–2830.

Michael Skinner. 2018. Product Categorization with LSTMs and Balanced Pooling Views. In *SIGIR 2018 Workshop on eCommerce (ECOM 18)*, SIGIR '18, Ann Arbor, MI, USA. ACM.

Nitish Srivastava, Geoffrey Hinton, Alex Krizhevsky, Ilya Sutskever, and Ruslan Salakhutdinov. 2014. Dropout: A Simple Way to Prevent Neural Networks from Overfitting. *Journal of Machine Learning Research*, 15:1929–1958.

Ilya Sutskever, Oriol Vinyals, and Quoc V Le. 2014. Sequence to Sequence Learning with Neural Networks. In Zoubin Ghahramani, Max Welling, Corinna Cortes, Neil D. Lawrence, and Kilian Q. Weinberger, editors, *Advances in Neural Information Processing Systems*, volume 27, pages 3104–3112. Curran Associates, Inc., Montréal, Canada.

Davy Weissenbacher, Abeed Sarker, Michael Paul, and Graciela Gonzalez-Hernandez. 2018. Overview of the Third Social Media Mining for Health (SMM4H) Shared Tasks at EMNLP 2018. In *Proceedings of the 3rd Workshop Social Media Mining for Health Applications (SMM4H)*, Brussels, Brussels. Association for Computational Linguistics.

Zichao Yang, Diyi Yang, Chris Dyer, Xiaodong He, Alex Smola, and Eduard Hovy. 2016. Hierarchical Attention Networks for Document Classification. In *Proceedings of the 2016 Conference of the North American Chapter of the Association for Computational Linguistics: Human Language Technologies*, pages 1480–1489, San Diego, CA, USA. Association for Computational Linguistics.

***Shot Or Not*:
Comparison of NLP Approaches for Vaccination Behaviour Detection

Aditya Joshi[1]   Xiang Dai[1,2]   Sarvnaz Karimi[1]
Ross Sparks[1]   Cécile Paris[1]   C Raina MacIntyre[3]

[1]CSIRO Data61, Sydney, Australia, [2]University of Sydney, Sydney, Australia
[3]The University of New South Wales, Sydney, Australia
{aditya.joshi,dai.dai,sarvnaz.karimi}@csiro.au
{ross.sparks,cecile.paris}@csiro.au, r.macintyre@unsw.edu.au

Abstract

Vaccination behaviour detection deals with predicting whether or not a person received/was about to receive a vaccine. We present our submission for vaccination behaviour detection shared task at the SMM4H workshop. Our findings are based on three prevalent text classification approaches: rule-based, statistical and deep learning-based. Our final submissions are: (1) an ensemble of statistical classifiers with task-specific features derived using lexicons, language processing tools and word embeddings; and, (2) a LSTM classifier with pre-trained language models.

1   Introduction

Public opinion about vaccines is diverse. Most people support vaccination, but some of these people do not receive vaccination. On the other hand, people who are vaccinated may also have concerns regarding the safety or efficacy of vaccines. In other words, a person's stance towards vaccines (referred to as '*vaccine hesitancy*') is distinct from whether or not they received a vaccine shot (referred to as '*vaccination behaviour*'). While automatic detection of vaccine hesitancy has been explored in the past, computational approaches to detect vaccination behaviour have been limited. Towards this, our paper deals with vaccination behaviour detection (SMM4H shared task #4). Vaccination behaviour and vaccine hesitancy can together help to understand penetration of vaccination programmes and the trust that communities place in large-scale vaccination programmes (Holt et al., 2016).

Vaccination behaviour detection is the task of predicting whether or not a given piece of text refers to a person receiving or intending to receive a vaccine. For example, the tweet '*I took the vaccine this morning, feeling great!*' is positive because the speaker reports having received the vac-

cine. On the contrary, '*Vaccines drastically reduce risks of infection*' is negative because the tweet describes vaccines but does not report a vaccine being administered.

Past work in vaccination behaviour detection uses n-grams as features of a statistical classifier (Skeppstedt et al., 2017; Huang et al., 2017). However, alternatives to n-grams have shown promise in several Natural Language Processing (NLP) tasks. Therefore, we compare three typical NLP approaches for vaccination behaviour detection: rule-based, statistical and deep learning techniques. Our submissions to the shared task use statistical and deep learning-based text classification. The systems are trained on a concatenation of the training and the validation set. The work reported in this paper ranked first among nine teams, as communicated by the shared task committee.

2   Approaches

In this section, we describe the three approaches that we employ for vaccination behaviour detection: Statistical, rule-based and deep learning-based.

2.1   Statistical Approach

Our statistical approach uses an ensemble of three classifiers: logistic regression, support vector machine with both using LIBLINEAR (Fan et al., 2008), and random forest using scikit-learn (Pedregosa et al., 2011). We use the following non-default parameters: (a) Positive misclassification cost is set to 3 in logistic regression; (b) 100 estimators in random forest. Majority voting is used to combine predictions from the classifiers, *i.e.*, a tweet must be predicted as positive by at least two classifiers for it to be predicted as positive by the ensemble.

The random forest classifier uses unigrams as features. The features for logistic regression and

Proceedings of the 3rd Social Media Mining for Health Applications (SMM4H) Workshop & Shared Task, pages 43–47
Brussels, Belgium, October 31, 2018. ©2018 Association for Computational Linguistics

Feature	Description	Type
N-grams	Unigrams and bigrams in the tweet	Boolean
Special Characters	@ and # which indicate user mentions and hashtags, ! and ?	Boolean
POS	Number of words of each POS tag	Count
Negation	Presence of negation words	Count
Word Similarity	Maximum value of similarity of words in the tweet and words indicating administration/reception of a vaccine	Real
Sentence Vector	Average of word vectors of the words in the tweet	Real
Length	Number of characters and words	Count
Emotion	Number of words of each emotion category	Count

Table 1: Features of the statistical approach.

support vector machine are summarised in Table 1. These features are:

1. **Uni/Bigrams**: Boolean;

2. **Special Characters**: A feature each indicating four special characters ?, #, @, !

3. **POS**: Count of each POS tag using NLTK POS tagger (Bird and Loper, 2004). This feature follows the intuition that presence of certain POS tags such as verbs may serve as signals;

4. **Negation**: Presence of a negation word. This is to serve as a negation feature where, although the act of receiving a vaccine is mentioned, the negation word changes the output class;

5. **Word Similarity**: For each word, we obtain similarity with '*receive*, '*get*' and '*take*', and use the highest similarity as this feature. We use pre-trained embeddings from Mikolov et al. (2013). This is to allow presence of words related to the act of receiving to be used as a signal for prediction;

6. **Sentence Vector**: A sentence vector is computed as an average of word vectors using GloVe embeddings (Pennington et al., 2014);

7. **Length**: Number of characters and words;

8. **Emotion**: Word counts of each emotion category as given by SenticNet (Cambria et al., 2014).

The combination of classifiers, misclassification costs and features has been experimentally validated.

2.2 Rule-based Approach

Since vaccination behaviour detection may appear to be only about detecting administration of a vaccine, we implement a naïve method to detect vaccination behaviour. Our rule-based approach looks for words indicating '*receive*' (without negation) to predict vaccination behaviour as follows:

1. If a tweet contains one among the words 'give', 'take', 'taking', 'gave', 'giving', 'get', 'getting', 'took', 'receive' or 'received' and no negation word, predict the tweet as positive.

2. Else, predict the tweet as negative.

2.3 Deep Learning-based Approach

We experiment with five typical deep learning-based models:

1. **Sentence Vector**: 200 dimensions; Logistic Regression. (*SV*)

2. **Dense Neural Network**: 64 dimensions, 1 inter. layer + 5 epochs (DNN)

3. **BiLSTM**: GloVe840B + 3 epochs + 50 lstm units + 0.25 dropout (*BiLSTM*)

4. **CNN**: GloVe840B + 5 epochs + 50 filters + 2 filter length + 0.75 dropout (CNN)

5. **LSTM-LM**: We pre-train a *language model* on the training dataset with a 3-layer LSTM. We then build a softmax layer on top of this pretrained LSTM, and fine-tune the neural network with supervision (Howard and Ruder, 2018).

All models are implemented using Tensor-Flow (Abadi et al., 2016). The parameters are experimentally determined.

Approach	F-Score	Accuracy
Skeppstedt et al. (2017)	76.84	87.01
Huang et al. (2017)	77.64	87.65
Statistical	80.75	**88.97**
Rule-based	40.48	64.91
SV	77.87	87.39
DNN	78.74	87.66
BiLSTM	78.30	87.30
CNN	78.40	87.60
LSTM-LM	**80.87**	88.94

Table 2: 10-fold cross-validation results (%) on the training dataset.

3 Experimental Setup

The shared task provided three labeled datasets of tweets for evaluation: a training dataset (5751 tweets of which 1692 are positive), a validation dataset (1215 tweets of which 306 are positive) and a test dataset (161 tweets, labels undisclosed).

We re-implement two past works as baselines (Skeppstedt et al., 2017; Huang et al., 2017). The two baselines use n-grams as features of statistical classifier.

4 Results

We present our results in six parts. We first describe the performances on the training, validation and test sets. Then, to understand the components contributing to the performance, we perform additional evaluation: (a) impact of the size of the training set on the performance; (b) impact of data source from which language model is trained in case of the deep learning approach; and (b) impact of the features on the performance of the statistical approach. Finally, we present an analysis of errors made by our system.

4.1 Performance on Training Set

The performance of our methods using 10-fold cross-validation is shown in Table 2. The performance of the re-implementation of baselines are comparable to the original papers. The low values in case of the rule-based approach highlight that vaccination behaviour detection is not a trivial task of detecting words that indicate administration of a vaccine. The best F-scores are achieved by the statistical approach (80.75%) and LSTM-LM (80.87%). This is an improvement of 3-4% over the baseline.

Approach	F-Score	Accuracy
Statistical	86.06	85.71
LSTM-LM	88.74	89.44

Table 3: Performance (%) on the test dataset.

	Statistical	LSTM-LM
20%	73.59	77.69
40%	75.17	78.58
60%	79.26	78.95
80%	80.54	79.52
100%	81.56	80.43

Table 4: F-scores (%) of the two best-performing approaches for varying size of the training set.

4.2 Performance on Validation and Test Sets

The statistical approach achieves an average F-score of 81.56%, while LSTM-LM achieves 80.43% on the validation set. Similarly, the performance of our methods on the test dataset is in Table 3. We obtain a F-score of 86.06% with the statistical approach and 88.74% with the LSTM-LM on the test set of 161 instances.

4.3 Impact of Size of the Training Set

To analyse the impact of the training set size on the resultant performance, we show the F-scores for the two best approaches for varying sizes of the training set in Table 4. '20%' indicates that 20% of the training set was used to train the system while the validation set was used for evaluation. We observe that when training on a small size of labeled data, LSTM-LM performs much better than statistical model. This shows the benefit of transfer learning that it can utilize knowledge learned from unlabeled data to train a model with a small number of labeled instances.

4.4 Impact of Language Model Source in LSTM-LM

A pre-trained language model represents knowledge learned from source data that is applied to a classifier. To understand if the domain of this source data has an impact on the performance of the resultant classifier, we compare how effective different domains are for vaccination behaviour detection. We compare three datasets in Table 6. The SMM4H dataset is the training dataset for the task while WikiText-103 (Merity et al., 2016)

Feature	ΔF-score (%)
POS	1.16
Special characters	0.97
Negation	0.66
Word similarity	0.15
Sentence vector	0.20
Length	0.39
Emotion	0.33

Table 5: Degradation in F-scores (%) of the statistical approach when each of the features is removed.

Source data	# of tokens	F-score
WikiText-103	101M	80.84 ($\pm$ 0.37)
IMDB	17M	81.15 ($\pm$ 0.83)
SMM4H	884K	80.43 ($\pm$ 0.67)

Table 6: F-scores (%) of the LSTM-LM when language model is pretrained on different source data.

and IMDB (Maas et al., 2011) are datasets from wikipedia and a movie review corpus respectively. The latter are significantly larger than the SMM4H dataset. However, they only result in a marginally higher performance.

4.5 Impact of Features in the Statistical Approach

To understand how the features contribute to the statistical approach, we conduct ablation tests. The degradation in F-score when each of the features is removed is in Table 5. The positive values in all fields validate the value of the proposed features. The highest degradation is observed in case of POS-based features.

4.6 Error Analysis

We analyse incorrectly predicted instances from the validation set. About 50% of errors have first or second person pronouns. Nearly 44% of false

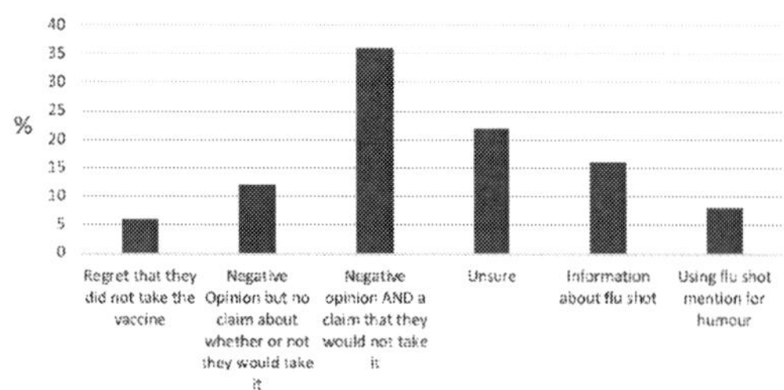

Figure 1: Sources of errors in false positives.

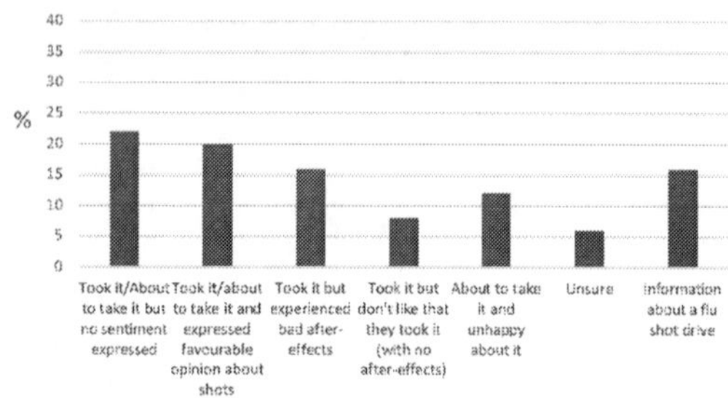

Figure 2: Sources of errors in false negatives.

negatives have negative sentiment about flu shots because of actual or expected, unpleasant side-effects. The ratio of false negatives to false positives is 1.40. An analysis of 50 random false positives and 50 random false negatives are shown in Figures 1 and 2 respectively. The label '*Unsure*' indicates that the error could not be assigned to any of the other categories. Some incorrectly classified instances for the different error sources are:

- Negative opinion but no claim whether they would take it, as in the case of '*Getting a flu vaccine after reading this article is crazy!*'.

- Mentions of taking a flu shot without expressing sentiment, such as '*Flu shots for hubby and daughter... check.*'.

- Took it or about to take it and expressed favourable opinion about shots, as in the case of the tweet '*We're headed to the @Brigham-Womens flu shot clinic! Getting vaccinated is good for you and your community.*'.

5 Conclusions

We evaluate three text classification approaches for the task of vaccination behaviour detection. The rule-based approach considers simple presence of words, the statistical approach uses an ensemble of classifiers and task-specific features while the deep learning approaches employ five neural models. On comparing the three approaches, we observe that an ensemble of statistical classifiers using task-specific features and a deep learning model using pre-trained language model and LSTM classifier obtain comparable performance for vaccination behaviour detection. Our findings in the error analysis which show that vaccine hesitancy often conflicts with vaccination behaviour detection, will be helpful for future work.

References

Martín Abadi, Paul Barham, Jianmin Chen, Zhifeng Chen, Andy Davis, Jeffrey Dean, Matthieu Devin, Sanjay Ghemawat, Geoffrey Irving, Michael Isard, et al. 2016. Tensorflow: A system for large-scale machine learning. In *OSDI*, volume 16, pages 265–283, Savannah, GA.

Steven Bird and Edward Loper. 2004. NLTK: The natural language toolkit. In *ACL*, page 31, Barcelona, Spain.

Erik Cambria, Daniel Olsher, and Dheeraj Rajagopal. 2014. SenticNet 3: A common and common-sense knowledge base for cognition-driven sentiment analysis. In *AAAI*, Quebec, Canada.

Rong-En Fan, Kai-Wei Chang, Cho-Jui Hsieh, Xiang-Rui Wang, and Chih-Jen Lin. 2008. LIBLINEAR: A library for large linear classification. *Journal of machine learning research*, 9(Aug):1871–1874.

D Holt, Fredric Bouder, C Elemuwa, G Gaedicke, A Khamesipour, B Kisler, S Kochhar, R Kutalek, W Maurer, P Obermeier, et al. 2016. The importance of the patient voice in vaccination and vaccine safetyare we listening? *Clinical Microbiology and Infection*, 22:146–153.

Jeremy Howard and Sebastian Ruder. 2018. Universal language model fine-tuning for text classification. In *Proceedings of the 56th Annual Meeting of the Association for Computational Linguistics (Volume 1: Long Papers)*, pages 328–339.

Xiaolei Huang, Michael C Smith, Michael J Paul, Dmytro Ryzhkov, Sandra C Quinn, David A Broniatowski, and Mark Dredze. 2017. Examining patterns of influenza vaccination in social media. In *AAAI Joint Workshop on Health Intelligence*, pages 542–546, San Francisco, CA.

Andrew L. Maas, Raymond E. Daly, Peter T. Pham, Dan Huang, Andrew Y. Ng, and Christopher Potts. 2011. Learning word vectors for sentiment analysis. In *Proceedings of the 49th Annual Meeting of the Association for Computational Linguistics: Human Language Technologies*, pages 142–150.

Stephen Merity, Caiming Xiong, James Bradbury, and Richard Socher. 2016. Pointer sentinel mixture models. *CoRR abs/1609.07843*.

Tomas Mikolov, Ilya Sutskever, Kai Chen, Greg S Corrado, and Jeff Dean. 2013. Distributed representations of words and phrases and their compositionality. In *Advances in Neural Information Processing Systems 26*, pages 3111–3119.

Fabian Pedregosa, Gaël Varoquaux, Alexandre Gramfort, Vincent Michel, Bertrand Thirion, Olivier Grisel, Mathieu Blondel, Peter Prettenhofer, Ron Weiss, Vincent Dubourg, et al. 2011. Scikit-learn: Machine learning in Python. *Journal of machine learning research*, 12(Oct):2825–2830.

Jeffrey Pennington, Richard Socher, and Christopher Manning. 2014. Glove: Global vectors for word representation. In *EMNLP*, pages 1532–1543, Doha, Qatar.

Maria Skeppstedt, Andreas Kerren, and Manfred Stede. 2017. Automatic detection of stance towards vaccination in online discussion forums. In *Proceedings of the International Workshop on Digital Disease Detection using Social Media*, pages 1–8, Taipei, Taiwan.

Neural DrugNet

Nishant Nikhil
IIT Kharagpur
Kharagpur India
nishantnikhil@iitkgp.ac.in

Shivansh Mundra
IIT Kharagpur
Kharagpur India
coolshivansh8@iitkgp.ac.in

Abstract

In this paper, we describe the system submitted for the shared task on Social Media Mining for Health Applications by the team Light. Previous works demonstrate that LSTMs have achieved remarkable performance in natural language processing tasks. We deploy an ensemble of two LSTM models. The first one is a pretrained language model appended with a classifier and takes words as input, while the second one is a LSTM model with an attention unit over it which takes character tri-gram as input. We call the ensemble of these two models: Neural-DrugNet. Our system ranks 2nd in the second shared task: Automatic classification of posts describing medication intake.

1 Introduction

In recent years, there has been a rapid growth in the usage of social media. People post their day-to-day happenings on regular basis. Weissenbacher et al. (2018) propose four tasks for detecting drug names, classifying medication intake, classifying adverse drug reaction and detecting vaccination behavior from tweets. We participated in the Task2 and Task4.

The major contribution of the work can be summarized as a neural network based on ensemble of two LSTM models which we call Neural DrugNet. We discuss our model in section 2. Section 3 contains the details about the experiments and preprocessing. In Section 4, we discuss the results and propose future works.

2 Model

Detection of drug-intake depends highly on:

- Whether the sentence conveys an intake.

- Whether a drug is mentioned in the sentence.

Long Short-Term Memory networks (Hochreiter and Schmidhuber, 1997) have been found efficient in tasks which need to learn structure of a sequential data. To learn a model which can value the first condition, we use LSTM based neural networks. Our first model is inspired from (Howard and Ruder, 2018), an LSTM model whose encoder is taken from a language model pre-trained on Wikipedia texts and fine-tuned on the tweets. After which a dense layer is used to classify into the different categories. And to also take into account the mentioning of a drug, which has not been there in the training data, we exploit the word structure of drug-names. Most of the drug-names have the same suffix. Example: melatonin, oxytocin and metformin have the suffix '-in'. We use a LSTM based model trained on the trigrams to learn that. Then, we take an ensemble of these two models. We give equal importance to both the models. That is, the prediction probability from Neural-DrugNet is the mean of the prediction probabilities from the two LSTM models. The predicted class is the one having the maximum prediction probability.

The training for the pre-trained language based LSTM model follows the guidelines given in the original paper (Howard and Ruder, 2018). They use discriminative fine-tuning, slanted triangular learning rates and gradual unfreezing of layers. For the character n-gram based LSTM model, as no fine-tuning is required, we train the model end-to-end.

3 Experiments

The data collection methods used to compile the dataset for the shared tasks are described in Weissenbacher et al. (2018).

Proceedings of the 3rd Social Media Mining for Health Applications (SMM4H) Workshop & Shared Task, pages 48–49
Brussels, Belgium, October 31, 2018. ©2018 Association for Computational Linguistics

3.1 Preprocessing

Before feeding the tweets to Neural-DrugNet, we use the same preprocessing scheme discussed in Nikhil et al. (2018). Then for the character trigram based model, we add an special character '$' as a delimiter for the word. That is, the character trigrams of 'ram' would be: '$ra', 'ram' 'am$'.

3.2 Results

We experimented with different type of architectures for both the tasks. Although any classifier like random forest, decision trees or gradient boosting classifier can be used. But due to lack of time, we used only support vector machine with linear kernel as baseline (denoted as LinearSVC). During development phase, we mistakenly used only accuracy as the metric for task2. The given results are based on a train-validation split of 4:1.

System	Accuracy
LinearSVC	0.675
LSTM with attention (words as input)	0.703
1D-CNN	0.651
Bi-LSTM with attention (words as input)	0.714
Bi-LSTM with attention (3-grams as input)	0.709
LSTM with encoder from Language Model	0.754
Neural DrugNet	**0.771**

Table 1: Results on validation data for Task2

System	F1-score
LinearSVC	0.751
Neural DrugNet	0.805
Neural DrugNet with LM fine-tuned on data from task3 also	**0.812**

Table 2: Results on validation data for Task4

The final results on best performing variant on test data for both the tasks are:

Precision	Recall	F1-score
0.520	0.491	0.505

Table 3: Results on test data for Task2

Accuracy	Precision	Recall	F1-score
0.857	0.824	0.897	0.859

Table 4: Results on test data for Task4

4 Conclusion and Future Work

In this paper, we present Neural-DrugNet for drug intake classification and detecting vaccination behavior. It is an ensemble of two LSTM models. The first one is a pretrained language model appended with a classifier and takes words as input, while the second one is a LSTM model with an attention unit over it which takes character tri-gram as input. It constantly outperforms the vanilla LSTM models and other baselines, which supports our claim that drug-intake classification and vaccination behavior detection rely on both the sentence structure and the character tri-gram based features. The performance reported in this paper could be further boosted by using a language model pretrained on tweets rather than the wikipedia texts. Furthermore, the ensemble module can be learned end-to-end by using a dense layer.

References

Sepp Hochreiter and Jürgen Schmidhuber. 1997. Long short-term memory. *Neural Comput.*, 9(8):1735–1780.

Jeremy Howard and Sebastian Ruder. 2018. Fine-tuned language models for text classification. *CoRR*, abs/1801.06146.

N. Nikhil, R. Pahwa, M. K. Nirala, and R. Khilnani. 2018. LSTMs with Attention for Aggression Detection. *ArXiv e-prints*.

Davy Weissenbacher, Abeed Sarker, Michael Paul, and Graciela Gonzalez-Hernandez. 2018. Overview of the third social media mining for health (smm4h) shared tasks at emnlp 2018. In *Proceedings of the 2018 Conference on Empirical Methods in Natural Language Processing (EMNLP)*.

IRISA at SMM4H 2018: Neural Network and Bagging for Tweet Classification

Anne-Lyse Minard[1]   Christian Raymond[1,2]   Vincent Claveau[1]
(1) CNRS, IRISA, Univ Rennes
(2) INSA Rennes, Rennes
Campus de Beaulieu, 35042 Rennes, France
`firstname.lastname@irisa.fr`

Abstract

This paper describes the systems developed by IRISA to participate to the four tasks of the SMM4H 2018 challenge. For these tweet classification tasks, we adopt a common approach based on recurrent neural networks (BiLSTM). Our main contributions are the use of certain features, the use of Bagging in order to deal with unbalanced datasets, and on the automatic selection of difficult examples. These techniques allow us to reach 91.4, 46.5, 47.8, 85.0 as F1-scores for Tasks 1 to 4.

1 Introduction

IRISA has participated in the four tasks of the SMM4H challenge (Weissenbacher et al., 2018). Yet, we have focused on Task 2 and 3, which are the most challenging ones, in particular because they have unbalanced data. Moreover, for Task 2, the three classes have very fuzzy boundaries, which makes some tweets difficult to classify even for humans. Our main contribution is to rely on Bagging (Bootstrap Aggregating) in order to deal with this problem of unbalanced data.

2 Methods

2.1 RNN: BiLSTM

For the four tasks, we have developed classifiers based on recurrent neural networks which consists in one Bidirectional LSTM layer (Graves et al., 2013) and a dense layer with a softmax activation as hidden layer. The input layer takes a representation of a tweet which consists in the word embeddings of each token and, depending of the task, a one-hot vector for each token or a one-hot vector for some medical terms in the tweet. Metamap Lite (Demner-Fushman et al., 2017) is used to extract specific medical terms from the tweets. We restrict the number of semantic types according to the task: for Task 1, we have selected only terms

related to drugs or substances; for Task 2, only to procedural terms; and for Task 3, we have selected both terms related to drugs and terms related to symptoms. For Task 1 and Task 2, we observe an improvement while using medical terms, whereas for Task 4 the use of metamap has no influence on the results. We use the word embeddings distributed by Grave et al. (2018). They have been trained with FastText (Bojanowski et al., 2017).

2.2 Bonzaiboost

During the development phase, we have used BONZAIBOOST, an implementation of the boosting algorithm adaboost.MH (Laurent et al., 2014) on decision trees. The results obtained are a bit lower than those of recurrent neural network methods. Yet, the experiments done with BONZAIBOOST allowed us to extract the most discriminating words, to choose the better features for the RNN, and to select the difficult examples (see Section 2.4). For Task 1, the important words found are drug names, such as *xanax*. For Task 2, the useful words are verbs indicating the action of taking a drug, the results of its intake, or the fact that a drug is needed (e.g. *took*, *need*). For Task 3, the discriminating words include symptom names (e.g. *dizzy*, *headache*). Finally for Task 4, no relevant discriminating words have been found. These findings help us to determine the semantic types of the medical terms to be used in the feature set.

2.3 Bagging

Bagging (Breiman, 1996) is a technique that consists in combining the prediction of different learners, where each "learner" uses only a sample of the original training set. We learn several models, with, for each, a subset of the training dataset, different training parameters (number of epochs, number of hidden layers...) and different feature sets. To deal with unbalanced datasets in Tasks

Proceedings of the 3rd Social Media Mining for Health Applications (SMM4H) Workshop & Shared Task, pages 50–51
Brussels, Belgium, October 31, 2018. ©2018 Association for Computational Linguistics

| Task | Run | Model | | | Input | | Data | F1 |
		BiLSTM	Bagging	epochs	metamap	tokens	cleaned	
T1	R1	x		3	x			**90.2**
	R2	x		3	x		x	90.0
	R3	x		3		x		88.1
T2	R1	x		3	x			67.9
	R2		x	3	x	x		67.7
	R3		x	1 to 5	x	x		**68.1**
T3	R1	x		3		x	x	50.1
	R2		x	3		x		**52.0**
	R3		x	1 to 7	x	x		46.6
T4	R1	x		3				87.7
	R2	x		3			x	**92.2**
	R3	x		3	x			87.2

Table 1: Description of the submitted runs and results obtained on the training dataset.

2 and 3, for Task 2 an equal number of instances of each class are chosen (2000 examples) and, for Task 3, every positive example is selected, and 20% of the negative ones are randomly selected.

Bagging seems to improve the results, especially because it allows the RNN to deal with more balanced datasets. For Task 1 and Task 4, bagging does not improve the results; this may be due to the results already being high (F1 $>$ 0.90), and for Task 1, to the data being already balanced.

2.4 Automatic cleaning of the datasets

Every manually annotated dataset may contain annotation errors or uncertain annotation due to the difficulty of the task. In order to improve the classification performance of our system, we have tried to clean up the training data by removing potential errors. We have considered that the tweets to be removed are those incorrectly classified although it was part of the training data used to train the model. More precisely, we proceed as follows: a model is trained on the complete training dataset; this model is then applied to predict the class of every example of this training dataset; the misclassified tweets are finally removed from the data; the cleaned dataset is then used to train the final model. We have removed 234, 183 and 250 examples respectively for Tasks 1, 3 and 4. For the Task 2, we have not observed improvement while removing difficult examples.

3 Evaluation

For each experiment the data is split into train set (80%) and dev set (20%). Evaluation is performed with a 5-fold cross validation, except when us-

ing bagging techniques. For the experiments with bagging, we train 20 models (with more models we do not get any improvement). The description of all the submitted runs and the obtained results on the training data are given in Table 1.

The official results are given in Table 2. The use of bagging techniques enables us to improve from 1.9 to 3.9 points the performance of our systems for Task 2 and Task 3. The automatic cleaning of the datasets is also a reason for a light improvement for Task 1 and Task 4.

Task	Run 1	Run 2	Run 3
T1	91.1	**91.4**	90.6
T2	43.6	45.5	**46.5**
T3	43.9	46.2	**47.8**
T4	84.4	**85.0**	82.4

Table 2: Final results in terms of F1-score.

References

Piotr Bojanowski, Edouard Grave, Armand Joulin, and Tomas Mikolov. 2017. Enriching Word Vectors with Subword Information. *Transactions of the Association for Computational Linguistics*, 5:135–146.

Leo Breiman. 1996. Bagging predictors. *Mach. Learn.*, 24(2):123–140.

Dina Demner-Fushman, Willie J Rogers, and Alan R Aronson. 2017. Metamap lite: an evaluation of a new java implementation of metamap. *Journal of the American Medical Informatics Association*, 24(4):841–844.

Edouard Grave, Piotr Bojanowski, Prakhar Gupta, Armand Joulin, and Tomas Mikolov. 2018. Learning Word Vectors for 157 Languages. In *Proceedings of the International Conference on Language Resources and Evaluation (LREC 2018)*.

Alex Graves, Abdel-rahman Mohamed, and Geoffrey Hinton. 2013. Speech recognition with deep recurrent neural networks. In *2013 IEEE International Conference on Acoustics, Speech and Signal Processing*, pages 6645–6649. IEEE.

Antoine Laurent, Nathalie Camelin, and Christian Raymond. 2014. Boosting bonsai trees for efficient features combination : application to speaker role identification. In *InterSpeech*, Singapour.

Davy Weissenbacher, Abeed Sarker, Michael Paul, and Graciela Gonzalez-Hernandez. 2018. Overview of the Third Social Media Mining for Health (SMM4H) Shared Tasks at EMNLP 2018. In *Proceedings of the 2018 Conference on Empirical Methods in Natural Language Processing (EMNLP), 2018*.

Drug-use Identification from Tweets with Word and Character N-grams

Çağrı Çöltekin
Department of Linguistics
University of Tübingen, Germany
ccoltekin@sfs.uni-tuebingen.de

Taraka Rama
Department of Informatics
University of Oslo, Norway
tarakark@ifi.uio.no

Abstract

This paper describes our systems in social media mining for health applications (SMM4H) shared task. We participated in all four tracks of the shared task using linear models with a combination of character and word n-gram features. We did not use any external data or domain specific information. The resulting systems achieved above-average scores among other participating systems, with F_1-scores of 91.22, 46.8, 42.4, and 85.53 on tasks 1, 2, 3, and 4 respectively.

1 Introduction

The increasing use of social media platforms world wide offers an interesting application of natural language processing tools for monitoring public health and health-related events on the social media. The social media mining for health applications (SMM4H) shared task (Weissenbacher et al., 2018) hosts four tasks aiming to identify mentions of different aspects medication use on Twitter. Briefly, the tasks and their descriptions are:

Task 1: Automatic detection of posts mentioning drug names.

Task 2: Automatic classification of posts describing medication intake.

Task 3: Automatic classification of adverse drug reaction mentioning posts.

Task 4: Automatic detection of posts mentioning vaccination behavior.

All tasks, except Task 2 are binary classification tasks. Task 2 requires three-way classification, including an uncertain class indicating posts mentioning possible medication intake.

For all tasks, we used linear SVM classifiers with character and word bag-of-n-gram features. We also experimented with other methods, including deep learning methods with gated RNNs

for building document representations. However, SVM models achieved best results on the development data. As a result, we only submitted results using linear SVMs, and we will only describe and discuss results of these model in this paper.

2 Methods and Experimental Procedure

We use the same general model for all tasks: linear SVM classifiers with character and word bag-of-n-gram features. Tokenization was done using a simple regular expression tokenizer that splits the text into consecutive alphanumeric and non-space, non-alphanumeric tokens. For each text to be classified, we extracted both character and word n-grams of order one up to a certain upper limit (specified below). All features are combined in a flat manner as a single text-feature matrix. We experimented with two feature weighting methods: tf-idf (Jurafsky and Martin, 2009, p.805) and BM25 (Robertson et al., 2009). The weighted features are then used for training an SVM classifier. We used one-vs-rest multi-class strategy when training the SVM classifier for task 2. All models were implemented in Python, using scikit-learn machine learning library (Pedregosa et al., 2011). The models are similar to the models we used in a few other text classification tasks (Çöltekin and Rama, 2018; Rama and Çöltekin, 2017; Çöltekin and Rama, 2017), where the models are explained in detail.

We tuned the models for each task separately, changing the maximum order of character and word n-gram features, case normalization, and SVM margin parameter 'C'. The parameter ranges explored during tuning was 0–12 for maximum character n-gram order, 0–7 for maximum word n-gram order, and 0.1–2.0 with steps of 0.1 for 'C'. We used 5-fold cross validation during tuning, using random search through the space of hy-

Proceedings of the 3rd Social Media Mining for Health Applications (SMM4H) Workshop & Shared Task, pages 52–53
Brussels, Belgium, October 31, 2018. ©2018 Association for Computational Linguistics

Task	tf-idf		BM25	
	devel.	test	devel.	test
1	90.17	90.87	90.13	91.22
2	76.42	46.8	76.45	46.5
3	93.52	40.4	93.42	42.4
4 (train)	89.22	–	89.41	–
4 (full)	90.16	85.53	90.16	85.53

Table 1: F1-scores of tf-idf and BM25 weighted models on the development set and the official test set. The F1-scores for task 2 are micro-averaged. The two set of scores for Task 4 reflect the difference between the full labeled-data set (including additional 1211 training instances) in comparison to the original training set.

perparameters described above. Approximately 1000 random hyperparameter settings were tried for each model. The models with the best parameter settings were retrained using the complete training data for producing the final predictions.

The source of the texts for all tasks is Twitter. At the time we downloaded them, some tweets were not available, resulting in training set sizes of 9182, 15 723, 16 888, and 5759 for tasks 1, 2, 3 and 4 respectively. Some of these numbers are substantially lower than that of intended number of training samples of 10 000, 17 000, 25 000, and 8180 respectively. For task 4, we also used an additional 1211 tweets, initially planned as the test set for this task. The test sets contained (approximately) 5000 tweets for tasks 1, 2 and 3, and a considerably smaller number (161) for task 4. All training sets showed some degree of class imbalance. The imbalance was particularly strong for tasks 3 and 4, where over 70 % and 90 % of the instances belonged to the negative class, respectively. Further information on the data sets can be found in Weissenbacher et al. (2018).

3 Results and Discussion

Table 1 presents F_1-scores of the models on each task. In general, we do not observe substantial differences between the term weighting schemes, but for some tasks the gap between training and development set scores is rather large. We do not know the system rankings at the time of writing, but only know that the results above are above the mean of the best-scores from all participating teams.

The systems we used for the shared task are simple, yet, effective classifiers with character and word n-gram features. The big discrepancies between the development and test set scores in task 2

and task 3 points either some differences between the distributions of the training and test sets, or it may also be due to large amount of missing tweets in our training set, indicating more data is likely to be particularly useful in these tasks. We also compared the effectiveness of two feature weighting systems, tf-idf and BM25, which did not show any substantial differences. Since our models were originally intended as baseline models, the scores presented in Table 1 were obtained without the use of any external data or source of information. Better results are likely by use of external information, such as appropriate dictionaries, term lists, or embeddings trained on large amounts of unlabeled data.

References

Çağrı Çöltekin and Taraka Rama. 2017. Tübingen system in vardial 2017 shared task: experiments with language identification and cross-lingual parsing. In *Proceedings of the Fourth Workshop on NLP for Similar Languages, Varieties and Dialects (VarDial)*, pages 146–155, Valencia, Spain.

Çağrı Çöltekin and Taraka Rama. 2018. Tübingen-Oslo at SemEval-2018 task 2: SVMs perform better than RNNs at emoji prediction. In *Proceedings of the 12th International Workshop on Semantic Evaluation (SemEval-2018)*, New Orleans, LA, United States.

Daniel Jurafsky and James H. Martin. 2009. *Speech and Language Processing: An Introduction to Natural Language Processing, Computational Linguistics, and Speech Recognition*, second edition. Pearson Prentice Hall.

F. Pedregosa, G. Varoquaux, A. Gramfort, V. Michel, B. Thirion, O. Grisel, M. Blondel, P. Prettenhofer, R. Weiss, V. Dubourg, J. Vanderplas, A. Passos, D. Cournapeau, M. Brucher, M. Perrot, and E. Duchesnay. 2011. Scikit-learn: Machine learning in Python. *Journal of Machine Learning Research*, 12:2825–2830.

Taraka Rama and Çağrı Çöltekin. 2017. Fewer features perform well at native language identification task. In *Proceedings of the 12th Workshop on Innovative Use of NLP for Building Educational Applications*, pages 255–260, Copenhagen, Denmark.

Stephen Robertson, Hugo Zaragoza, et al. 2009. The probabilistic relevance framework: BM25 and beyond. *Foundations and Trends® in Information Retrieval*, 3(4):333–389.

Davy Weissenbacher, Abeed Sarker, Michael Paul, and Graciela Gonzalez-Hernandez. 2018. Overview of the third social media mining for health (smm4h) shared tasks at EMNLP 2018. In *Proceedings of the 2018 Conference on Empirical Methods in Natural Language Processing (EMNLP)*.

Automatic Identification of Drugs and Adverse Drug Reaction Related Tweets

Segun Taofeek Aroyehun
CIC, Instituto Politécnico Nacional
Mexico City, Mexico
`aroyehun.segun@gmail.com`

Alexander Gelbukh
CIC, Instituto Politécnico Nacional
Mexico City, Mexico
`www.gelbukh.com`

Abstract

We describe our submissions to the Third Social Media Mining for Health Applications Shared Task. We participated in two tasks (tasks 1 and 3). For both tasks, we experimented with a traditional machine learning model (Naive Bayes Support Vector Machine (NBSVM)), deep learning models (Convolutional Neural Networks (CNN), Long Short-Term Memory (LSTM), and Bidirectional LSTM (BiLSTM)), and the combination of deep learning model with SVM. We observed that the NBSVM reaches superior performance on both tasks on our development split of the training data sets. Official result for task 1 based on the blind evaluation data shows that the predictions of the NBSVM achieved our team's best F-score of 0.910 which is above the average score received by all submissions to the task. On task 3, the combination of of BiLSTM and SVM gives our best F-score for the positive class of 0.394.

1 Introduction

The emergence of social media platforms such as Twitter has led to the availability of huge amount of data for research purposes. Public health monitoring using this non-traditional mode of communication has received attention in recent times. The third edition of Social Media Mining for Health Applications (SMM4H) (Davy et al., 2018) shared task aims to facilitate pharmacovigilance research using social media data.

We participated in tasks 1 and 3. The purpose of task 1 is to identify tweets that contain drug name(s) while task 3 focuses on recognizing Twitter posts mentioning adverse drug reaction (ADR). Both tasks are binary classification tasks. The evaluation metrics for both tasks are the precision, recall, and F1 scores of the positive class.

In the following sections, we describe the data, our approach, results, and conclusion.

Task	Train set		Test set
	neg class	pos class	
1	4356	4700	5382
3	15326	1351	5000

Table 1: Number of Examples in the Train and Test Sets for Tasks 1 and 3

2 Data

The shared task organizers provided datasets consisting of tweet IDs and their corresponding label as well as a script to download the text of the tweets. Using the IDs, textual data was gathered from Twitter. For task 1, the tweets were annotated for the presence of at least one mention of drug name. The presence of ADR mention was equally annotated for task 3. We downloaded a total of 9056 tweets out of 9625 expected tweets as training data for task 1. Also, 16677 tweets were retrieved out of 25630 expected tweets for task3. Table 1 shows the number of examples per label in the training data for task 1 and task 3. For task 1, the number of examples per class is almost equal. For task 3, the number of examples per label are highly imbalanced with almost 92% of the examples belonging to the negative class (non-ADR) and approximately 8% of the training data are of the positive class (ADR). The blind test set consists of 5382 tweets and 5000 tweets for task 1 and task 3 respectively. We cleaned the datasets by removing special and repeated characters, numbers, URL, and hashtags. To handle mispellings, we ran a spell checker.

3 Method

Our approach to both tasks 1 and 3 is very similar. We experimented with NBSVM, deep learning models, and the combination of a deep learning model as feature extractor and SVM as classifier.

Proceedings of the 3rd Social Media Mining for Health Applications (SMM4H) Workshop & Shared Task, pages 54–55
Brussels, Belgium, October 31, 2018. ©2018 Association for Computational Linguistics

Task	Classifiers			
	NBSVM	CNN	LSTM	BiLSTM
1	**0.909**	0.877 (0.888)	0.848 (0.781)	0.876 (0.798)
3	**0.624**	0.619 (0.549)	0.591 (0.391)	0.622 (0.321)

Table 2: F1 Score of the Positive Class on our Development Split of the Training set using NBSVM and Deep Learning Models (For the deep learning models, the scores are the average of three runs and the values in parenthesis are for the corresponding character level model)

NBSVM is a strong baseline (Wang and Manning, 2012). The choice of the deep learning model to use as feature extractor was informed by the average performance across three runs on our development split. The train-development split used for task 1 is 90% for training and 10% for development. For task 3, the development split was generated after random undersampling of the majority class. We maintained class imbalance in the ratio 1:3 of the minority class to the majority class. As shown in Table 2, the best performing deep learning model for task 1 was CNN and BiLSTM for task 3.

In our experiments, the NBSVM model uses the log-count ratios over character n-grams ranging from 1 to 5 characters as features. In the deep learning models, we employed the pre-trained fastText word embedding [1]. The SVM model was trained using the RBF kernel.

For the deep learning models, we used the binary cross entropy loss function as our objective function. To optimize the loss function through backpropagation, we used ADAM optimizer with learning rate of 0.001. We ran the models for 100 epochs with earlystopping and dropout layers with probability of 0.2 in order to avoid overfitting.

4 Results

Table 3 shows the performance of our systems on the task 1 evaluation data. The NBSVM model achieved our best recall (0.899) and F1 (0.910) scores. These scores are above average. The average precsion, recall, and F1 scores are 0.8904, 0.872, and 0.880 respectively. The CNN model was marginally higher than the NBSVM by 0.002 on the precision score. For task 3, Table 4 shows that our BiLSTM+SVM model is our best submission reaching our best score on precision (0.314) and F1 (0.394) scores for the ADR class. The NB-

System	P	R	F
NBSVM	0.920	**0.899**	**0.910**
CNN	**0.922**	0.786	0.848
CNN+SVM	0.909	0.803	0.853

Table 3: Scores on the Evaluation Data for Task 1 (P-Precision; R-Recall; F-F1 measure)

System	P	R	F
NBSVM	0.258	**0.795**	0.390
BiLSTM	0.293	0.586	0.390
BiLSTM+SVM	**0.314**	0.529	**0.394**

Table 4: Scores on the Evaluation Data for Task 3 (P-Precision for the ADR class; R-Recall for the ADR class; F-F1 measure for the ADR class)

SVM model achieves a better recall on the ADR class, 0.795. The difference in recall scores suggests that an ensemble of classifiers might lead to a better F1 score.

5 Conclusion

In this paper, we describe our participation in tasks 1 and 3 of the SMM4H shared tasks. We developed three classifiers for both tasks using NB-SVM, deep learning models (CNN, LSTM, and BiLSTM), and the comboination of a deep learning model and SVM. For task 1, we achieved our best submission using the NBSVM. The BiLSTM+SVM model achieved our best F1 score for the ADR class on task 3 while the NBSVM model scores better in terms of recall.

As future direction, we would like to investigate the use of informed sampling techniques in handling class imbalance. Also, we will explore the enrichment of the training data with semantic and conceptual domain knowledge that could provide relevant priors for the classifiers.

References

Weissenbacher Davy, Sarker Abeed, Paul Michael, and Gonzalez-Hernandez Graciela. 2018. Overview of the third social media mining for health (smm4h) shared tasks at emnlp 2018. In *Proceedings of the 2018 Conference on Empirical Methods in Natural Language Processing (EMNLP)*.

Sida Wang and Christopher D Manning. 2012. Baselines and bigrams: Simple, good sentiment and topic classification. In *Proceedings of the 50th Annual Meeting of the Association for Computational Linguistics: Short Papers-Volume 2*, pages 90–94. Association for Computational Linguistics.

---

[1] `https://s3-us-west-1.amazonaws.com/fasttext-vectors/wiki.en.zip`

UZH@SMM4H: System Descriptions

Tilia Ellendorff,[*] **Joseph Cornelius**,[†] **Heath Gordon**,[*] **Nicola Colic**,[*] **Fabio Rinaldi**[*]
[*]Institute of Computational Linguistics, University of Zurich (`{name.surname}@uzh.ch`)
[†]Institute of Neuroinformatics, University of Zurich and ETH (`jocorn@ini.phys.ethz.ch`)

Abstract

Our team at the University of Zürich participated in the first 3 of the 4 sub-tasks at the Social Media Mining for Health Applications (SMM4H) shared task. We experimented with different approaches for text classification, namely traditional feature-based classifiers (Logistic Regression and Support Vector Machines), shallow neural networks, RCNNs, and CNNs. This system description paper provides details regarding the different system architectures and the achieved results.

1 Introduction

The 2018 edition of the Social Media Mining for Health Applications (SMM4H) challenge (Weissenbacher et al., 2018) consists of 4 tasks, all of which can be framed as document classification problems: automatic detection of posts mentioning a drug name (task 1), automatic classification of posts describing medication intake (task 2), automatic classification of posts mentioning adverse drug reaction (task 3) and vaccination behavior, respectively (task 4). Our team participated in the first three of them.

While tasks 1 (*drug name detection*) and 3 (*adverse drug reaction mentioning detection*; ADR) consisted in binary text classification, task 2 (*medication intake classification*) included three classes: *personal medication intake*, *possible medication intake* and *non-intake*.

2 Data Description and Pre-processing

For each task, participants were provided with a dataset. The tweets were provided by ID, and had to be downloaded individually from Twitter. Because of that, not all tweets in the datasets were available anymore at the time of our participation. The number of tweets that we had at our disposal are shown in Table 1.

Task	Labels	Tweets	Train	Develop.
1	0	4462	4357	105
	1	4776	4657	119
Total		9238	9014	224
2	1	3198	3129	69
	2	5162	5058	104
	3	7155	7028	127
Total		15515	15215	300
3	0	15416	15148	268
	1	1359	1327	32
Total		16775	16475	300

Table 1: Overview of available tweets for each task, split by us into a training and a small development set.

We tokenized the tweets using SpaCy (Honnibal and Montani, 2017) and applied a number of pre-processing steps before further processing the tweets. These include the following:

- We found that URLs are frequently merged with the preceding token. Therefore, we split before URLs (i.e. before "http").
- We split URLs into their components (i.e. parts of each URL are treated as separate tokens).
- We split all tokens at camel case (e.g. "MedicationProblems" is split into "Medication" and "Problems").
- We stripped the hashtag symbol (#) from all tokens where it applies.
- We replaced "w/" and "w/o" by their full versions ("with" and "without").
- We additionally split at the following punctuation symbols: -/.]
- We replaced numbers and usernames by placeholder tokens.

Following these pre-processing steps, we used SpaCy for lemmatization and part-of-speech tagging.

Proceedings of the 3rd Social Media Mining for Health Applications (SMM4H) Workshop & Shared Task, pages 56–60
Brussels, Belgium, October 31, 2018. ©2018 Association for Computational Linguistics

	Task 1			Task 2			Task 3		
	P	R	F	P	R	F	P	R	F
Logreg:	0.861	0.861	0.861	0.591	0.565	0.578	0.917	0.344	0.500
MLP:	0.861	0.861	0.861	0.679	0.522	0.590	0.750	0.281	0.409
Lin SVM:	0.534	0.534	0.534	0.575	0.609	0.592	0.571	0.375	0.453
Shallow NN:	0.577	0.381	0.459	0.780	0.565	0.655	0.550	0.344	0.423
RCNN:	0.916	0.924	0.920	0.468	0.638	0.540	0.185	0.156	0.169

Table 2: Results of feature based systems on the development set. For task 2, results are micro-averaged over the two positive labels only. For task 3, they are micro-averaged over the positive label only.

3 System Descriptions

The following sections give an overview of the system architectures with which we experimented. The results obtained on our development set by a selection of configurations (as described below) are shown in Table 2.

3.1 Feature-based systems

The feature-based classifiers include the following features:

- bag-of-lemmas (unigrams and bigrams)
- averaged pre-trained word embeddings (Sarker and Gonzalez, 2017)
- a binary feature providing information if the tweet contains any token found in a terminology list described later in this section
- a set of features recording all exact matches from that list as observed in the training data. This allows the classifier to assign a weight depending on the number of positive or negative tweets in which these terms are observed.

We experimented with Naive Bayes, linear SVMs, Logistic Regression and Multilayer Perceptron classifiers. However, since Naive Bayes consistently gave us the worst performance on the development set, we excluded it. Our Multilayer Perceptron has two hidden layers using tanh activation with 100 and 50 units, respectively, and applies an Adam optimizer with an adaptive learning rate.

To improve on our results, we employ two term lists; one with terms derived from an external resource, and one with terms extracted from the task data. Firstly, we use a manually curated list of drug names, derived from RX Norm (Nelson et al., 2011), which we had originally created for a different project. RX Norm is a normalized list of all clinical drugs available in the US, indexing them by commercial name and compounds. We

142	drug names
85	chemical compounds
42	class of drugs (such as analgesics)
41	misspellings
39	symptoms
38	related term (such as addictive)
17	hashtag (such as #advilsinuscrowd)
16	abbreviations (such as alka)
12	pharmaceutical company
73	others (plants, medical devices etc)
505	**total**

Table 3: Categories of terms derived from tweets.

compared this list with a list of the 10000 most common English words in order to determine the amount of ambiguity and found only a negligible overlap. This means that both chemical compounds and brand names are very specific in most cases and therefore only show a very small amount of ambiguity.

However, to better account for the fact that social media data is noisy, and users misspell and abbreviate drug names or use different names not contained in the vocabulary, we constructed a second list: We gathered tokens from positive tweets in the training data which do not contain any drug names from the list above. From this set, only tokens that do not occur in the 20000 most frequent natural language words as computed on Google Books Ngram Corpus were kept, and evaluated manually if they refer to a drug. This method revealed common misspellings such as *adderal* (instead of *adderall*) or *codrol* (instead of *codral*), but also lead to the identification of several word categories that can be positive predictors for drug usage, such as diseases and symptoms. Table 3 lists the categories of the terms extracted in this fashion.

3.2 Shallow Neural Network with Tunable Embeddings

This is a simple system based on end-to-end learning within a shallow neural network. The first layer consists of tunable pre-trained embeddings followed by average pooling and a dense layer with sigmoid activation to reach a final classification decision. The embedding layer uses the fasttext embeddings trained on the English version of Wikipedia (Bojanowski et al., 2016), which, during training, we fine-tune to the task. We use cross-entropy as loss function and Adam for optimization. Furthermore, we apply early stopping using a small portion of the training set.

3.3 Recurrent Convolutional Neural Network (RCNN)

For task 1, we additionally apply a combination of a recurrent neural network and a convolutional neural network. The recurrent convolutional neural network (RCNN) (Lai et al., 2015) uses recurrent structures which enables it to capture the context information of each word while simultaneously producing minimal noise. Additionally, it uses a max-pooling layer to capture the relevance of every word in the text. Our recurrent structure is a two layer stacked bidirectional network with gated recurrent unit (GRU) (Cho et al., 2014) cells. The final hidden states of the recurrent structure are the input to the 1D-max-pooling layer. The model is based on Prakash Pandey's implementation of Text Classification in PyTorch. We experimented with a character-based and a word-based version, which are described in the following.

The words of the lemmatized version of the tweets serve as input for the word-based RCNN. We have experimented with various word embeddings in combination with and without a lemmatized input. We used embeddings that were trained on a lemmatized corpus of tweets as well as embeddings that were trained on a non-lemmatized corpus of tweets. In the final version we used embeddings trained on a non-lemmatized corpus for processing the lemmatized version of corpus. Even if it might seems methodologically incorrect, it this was relatively simple to do, and it was empirically found to produce the best results on our development corpus. Our model has 256 hidden states for each direction of the bidirectional recurrent network. We use Adam with a learning rate of 0.00008 for the optimization and softmax

cross entropy to compute the loss. We utilize L2 regularization with a rate of 0.005 and to prevent overfitting, we employ a dropout with a rate of 0.8. The whole system learns for 45 epochs with a batch size of 256. We apply domain-specific, pretrained word embeddings (Sarker and Gonzalez, 2017) that are trained on 1 billion tweets from drug-related conversations on Twitter. In addition, we append one dimension to the word embeddings to determine whether a word is listed in the list of RX Norm drug names mentioned before. If a word is included in the list, we insert the value 5 and if it is not included we insert the value -1. In the case of drug names for which no pre-trained embedding is available, we generate a generic embedding vector by averaging the vectors of all RX terms for which an embedding can be found.

We also experimented with a character-based RCNN (using 1-grams and 3-grams) as another approach to the detection of drug name mentioning tweets. However, we did not use this system in the final submission since its performance was consistently worse than that of the word-based model. We considered a combination of the two approaches, however we could not implement it due to lack of time. As input we use the lemmatized version of the tweets converted to the specific character N-gram. Each N-gram corresponds to a unit that is fed into the RNN in the form of its embedding. The same hyperparameters as in the word-based RCNN are used in the character-based RCNN, except that we apply a learning rate of 0.0001, an L2 regularization with a rate of 0.001 and train the system only for 40 epochs. To test the model we have divided the given training set into a validation, test and training set[1]. With the test set, which contains 1000 samples, we achieve an precision of 0.8879, a recall of 0.8840 and an F1 score of 0.8860.

After the challenge, we performed a small error analysis on the results obtained by the word-based RCNN over the 244 tweets in the development set. We found that 19 tweets were misclassified, 10 tweets were mistakenly classified as drug name mentioning, but six of them contain words like 'vitamin', 'maca' and 'pills'. The distinction made by the dataset between the fine nuances of some substances as endogenous rather than as a supplement seems to be a problem for the system. Nine

---

[1]The partition of the data used within this set-up is different from the partitions reported in table 1

tweets were falsely identified as drug-free tweets, one of these contains a misspelled drug name and another one contains incorrectly tokenized words (e.g. *aspirin!*). Another two tweets contained the word "weed" in the medical context and the word "cannabinoid", which may have been classified as drug names rather than medication names. For the remaining tweets, there is no specific property that would lead to a misclassification.

For comparison, we briefly report on results obtained using a simple biLSTM (bidirectional long short term memory) model as a baseline. The biLSTM model has 128 hidden states for each direction of the bidirectional recurrent network. Adam is used for the optimization with a learning rate of 0.0001. As loss function we use softmax cross entropy. We apply L2 regularization with a rate of 0.005 and a dropout with a rate of 0.5 to achieve a greater generalization. We run the model for 30 epochs with a batch size of 128. On the development set with 1000 tweets the baseline biLSTM achieves an F1 score of 0.881, while the previously described final version of the RCNN achieves an F1 score of 0.92.

3.4 Ensemble of Convolutional Neural Networks (CNNs)

For task 3 we created a tiered ensemble of convolutional neural networks (CNNs). To create the first ensemble, we downsampled the majority class by splitting it up into 5 equally sized training sets. The minority class data remained unsampled and was paired with each of those majority class splits, giving 5 data sets of a 1:2 minority-majority ratio. A CNN was trained on each of these samples for 20 epochs. Input features are the pre-trained word embeddings by Sarker and Gonzalez (2017). The decisions of each CNN for each sample are fed into a simple voting classifier. Twenty of these ensembles were created, and their predictions are fed into a simple majority vote classifier, forming the final set of predictions. This system is based on the general architecture used by (Friedrichs et al., 2018) whereas the individual CNNs is based on (Kim, 2014). Despite promising results on our development set (F: 48.3), this setting performed poorly in the official evaluation (Run 3 of task 3), probably due to a configuration error.

		Precision	Recall	F-Score
Task 1	run 1	0.908	0.834	0.870
	run 2	0.927	0.878	0.902
	run 3	0.908	0.856	0.878
	Mean	**0.890**	**0.872**	**0.880**
Task 2	run 1	0.315	0.434	0.365
	run 2	0.371	0.437	0.401
	run 3	0.431	0.368	0.397
	Best	**0.654**	**0.783**	**0.713**
Task 3	run 1	0.593	0.231	0.333
	run 2	0.455	0.436	0.445
	run 3	0.132	0.935	0.232
	Best	**0.442**	**0.636**	**0.522**

Table 4: Official scores for our submissions, compared with scores obtained by other participating systems.

4 Results

Our official results are summarized in Table 4, compared with other official results as currently available to us (score means for task 1, scores of the system with best F-score for task 2 and 3). Below a brief description of our submitted runs.

For **task 1**, Run 1 was based on the best performing among the models described in section 3.1 (logistic regression). Run 2 was based on the word-based RCNN model described in 3.3. Run 3 was based on a rather crude attempt to improve the scores of a previous run based on a manually curated version of the drug list described in section 3.1, flipping a negative into a positive if the presence of one of terms in the list was detected. The idea was that such a presence should be considered as a highly reliable indicator that the tweet is positive. Experiments on the training corpus had shown generally an increase in recall with a minimal loss in precision. This was confirmed in the official results, where the correction described above was applied to the results of Run 1. However, after the competition we asked the organizers to score another run generated by applying the same "corrector" to Run 2, but in this case the results improved only slightly (P: 0.890, R: 0.872, F:0.880).

For **task 2**, Run 1 was based on the best performing among the models described in section 3.1 (linear SVM), Run 2 with the second best model (Multilayer Perceptron), Run 3 with the Shallow Neural Network model described in section 3.2.

For **task 3**, Run 1 was generated with the Logistic Regression model, Run 2 with the Shallow Neural Network model, Run 3 with the complex CNN ensemble approach described in section 3.4. However, we suspect a bug in the latter approach since the results were worse than anticipated.

5 Conclusion

In this system description paper we provide details and results for the different approaches with which we experimented for our participation in 3 sub-tasks of the SMM4H shared task. Our interest in this shared task stems from the fact that we are involved in a recently started research project where we will process social media data, including tweets in a related domain. Therefore we plan to continue our experiments with the datasets provided and report new results at the final workshop.

Acknowledgments

We gratefully acknowledge the support of the Swiss National Science Foundation (grants $CR30I1\_162758$ and 407440_167381) and Innosuisse (grant $25587.2PFES - ES$).

References

Piotr Bojanowski, Edouard Grave, Armand Joulin, and Tomas Mikolov. 2016. Enriching Word Vectors with Subword Information. *arXiv preprint arXiv:1607.04606*.

Kyunghyun Cho, Bart van Merriënboer, Çalar Gülçehre, Dzmitry Bahdanau, Fethi Bougares, Holger Schwenk, and Yoshua Bengio. 2014. Learning phrase representations using rnn encoder–decoder for statistical machine translation. In *Proceedings of the 2014 Conference on Empirical Methods in Natural Language Processing (EMNLP)*, pages 1724–1734, Doha, Qatar. Association for Computational Linguistics.

Jasper Friedrichs, Debanjan Mahata, and Shubham Gupta. 2018. InfyNLP at SMM4H Task 2: Stacked Ensemble of Shallow Convolutional Neural Networks for Identifying Personal Medication Intake from Twitter. *arXiv preprint arXiv:1803.07718*.

Matthew Honnibal and Ines Montani. 2017. SpaCy 2: Natural language understanding with Bloom embeddings, convolutional neural networks and incremental parsing. *To appear*.

Yoon Kim. 2014. Convolutional neural networks for sentence classification. *arXiv preprint arXiv:1408.5882*.

Siwei Lai, Liheng Xu, Kang Liu, and Jun Zhao. 2015. Recurrent Convolutional Neural Networks for Text Classification. *Twenty-Ninth AAAI Conference on Artificial Intelligence*, pages 2267–2273.

Stuart J Nelson, Kelly Zeng, John Kilbourne, Tammy Powell, and Robin Moore. 2011. Normalized names for clinical drugs: Rxnorm at 6 years. *Journal of the American Medical Informatics Association*, 18(4):441–448.

Abeed Sarker and Graciela Gonzalez. 2017. A corpus for mining drug-related knowledge from Twitter chatter: Language models and their utilities. *Data in Brief*, 10:122–131.

Davy Weissenbacher, Abeed Sarker, Michael Paul, and Graciela Gonzalez-Hernandez. 2018. Overview of the Third Social Media Mining for Health (SMM4H) Shared Tasks at EMNLP 2018. In *Proceedings of the 2018 Conference on Empirical Methods in Natural Language Processing (EMNLP)*.

Deep Learning for social media health text classification

T.Y.S.S.Santosh   Vaibhav Gambhir   Animesh Mukherjee
IIT Kharagpur
West Bengal – 721302
India
{santoshtyss,v.gambhir}@gmail.com
{animeshm}@cse.iitkgp.ac.in

Abstract

This paper describes the systems developed for 1st and 2nd tasks of the 3rd Social Media Mining for Health Applications Shared Task at EMNLP 2018. The first task focuses on automatic detection of posts mentioning a drug name or dietary supplement, a binary classification. The second task is about distinguishing the tweets that present personal medication intake, possible medication intake and non-intake. We performed extensive experiments with various classifiers like Logistic Regression, Random Forest, SVMs, Gradient Boosted Decision Trees (GBDT) and deep learning architectures such as Long Short-Term Memory Networks (LSTM), jointed Convolutional Neural Networks (CNN) and LSTM architecture, and attention based LSTM architecture both at word and character level. We have also explored using various pre-trained embeddings like Global Vectors for Word Representation (GloVe), Word2Vec and task-specific embeddings learned using CNN-LSTM and LSTMs.

1   Introduction

The tasks (Davy Weissenbacher, 2018) involve NLP challenges on social media mining for health monitoring and surveillance and in particular pharmaco-vigilance. This requires processing noisy, real-world, and substantially creative language expressions from social media. The proposed systems should be able to deal with many linguistic variations and semantic complexities in various ways people express medication-related concepts and outcomes. The tasks present several interesting challenges including the noisy nature of the data, the informal language of the user posts, misspellings, and data imbalance.

Deep learning has the potential to improve analysis of social media text because of its ability to learn patterns from unlabelled data (Arel et al., 2010). This property has enabled deep learn-ing to produce breakthroughs in the domain of image, text and speech recognition. Moreover, deep learning has the ability to generalize learnt patterns beyond data similar to the training data, which can be advantageous while dealing with social media text. Despite the breakthroughs brought by deep learning, improvements are still to be made to further optimise it and improve its performance (LeCun et al., 2015). This paper proposes to explore how the emerging advantages of deep learning can be expanded upon to address the pertinent challenges for social media text analysis.

For Task 1, tweets are required to be distinguished those that mention any drug names or dietary supplement. For Task 2, the data-set contains tweets mentioning a drug and the objective is to classify the tweet into three classes. The class descriptions are as follows: personal medication intake tweets in which the user clearly expresses a personal medication intake/consumption; possible medication intake tweets that are ambiguous but suggest that the user may have taken the medication; non-intake tweets that mention medication names but do not indicate personal intake.

2   Method

This section describes the deep learning architectures we used for the tasks, described as follows: 1) CNN-LSTM 2) LSTM with attention mechanism. The subsections give a brief description of both of the approaches.

2.1   CNN-LSTM

With the development of deep learning, typical deep learning models such as CNNs and recurrent neural networks (RNNs) have achieved remarkable results in computer vision and speech recognition. Word embeddings, CNNs (Kim, 2014) and RNNs (Graves, 2012) have been applied to text classification and got good results. CNN and RNN are two mainstream architectures for such model-

61

Proceedings of the 3rd Social Media Mining for Health Applications (SMM4H) Workshop & Shared Task, pages 61–64
Brussels, Belgium, October 31, 2018. ©2018 Association for Computational Linguistics

ing tasks, which adopt totally different ways of understanding natural languages. In this system, we combine the strengths of both architectures and use a novel and unified model called CNN-LSTM (Zhou et al., 2015) for sentence classification. CNN-LSTM utilizes CNN to extract a sequence of higher-level phrase representations, and are fed into an LSTM to obtain the sentence representation. We take the word embeddings as the input of our CNN model in which windows of different length and various weight matrices are applied to generate a number of feature maps. After convolution and pooling operations, the encoded feature maps are taken as the input to the LSTM model. The long-term dependencies learned by LSTM can be viewed as the sentence- level representation. The sentence-level representation is fed to the fully connected network and the softmax output reveals the classification result. The deep learning algorithm we put forward to use for these tasks differs from the existing methods in that our model takes advantage of the encoded local features extracted from the CNN model and the long-term dependencies captured by the LSTM model.

2.2 LSTM with attention mechanism

A limitation of the usual LSTM architecture is that it encodes the input sequence to a fixed length internal representation. This imposes limits on the length of input sequences that can be reasonably learned. A recently proposed method for easier modeling of long-term dependencies is attention. Attention mechanisms allow for a more direct dependence between the state of the model at different points in time. Attention-based RNNs have proven effective in a variety of sequence transduction tasks, including machine translation (Bahdanau et al., 2014), image captioning (Xu et al., 2015), and speech recognition (Chan et al., 2016). This is achieved by keeping the intermediate outputs from the LSTM from each step of the input sequence and training the model to learn to pay selective attention to these inputs and relate them to items in the output sequence.

3 Experiment

This section details how the proposed approach is applied to Task 1 and Task 2 data sets. Task 1 is a binary classification problem and task 2 is a multiclass classification problem. The dataset statistics are given in Table 1 and Table 2. The dataset for each task includes training data and test data.

As baselines, we experimented with several classifiers like Logistic Regression, Random Forest, SVMs, Gradient Boosted Decision Trees (GBDT). We have used TF-IDF to extract the feature values. We then used the CNN-LSTM and attention based LSTM networks and are trained (fine-tuned) using labeled data with backpropagation. We have also experimented with CNN-LSTM and attention based LSTM networks by using pre-trained embeddings such as GloVE and Word2vec for word level and we have also experimented them at character level. These networks also learn task-specific word embeddings. Therefore, for each of the networks, we also experimented by using these embeddings as features and trained various classifiers like Logistic Regression, Random Forest, SVMs, GBDT.

4 Results

We have submitted the top 3 systems for each task on validation data. Table 3 and 4 describes the precision, recall and F1-score on the validation data and test data for Task 1 respectively. We have selected top 3 based on cumulative score of recall, precision and F1-score. On test data character level LSTM-CNN gave the good precision and F1-score whereas word level LSTM with attention embeddings trained on Naive bayes classifier gave the good recall. Table 5 and 6 describes the precision, recall and F1-score on the validation data and test data for the Task 2. On test data character level LSTM-CNN gave highest micro-averaged precision, recall and F1-score.

5 Conclusion

In this paper we described briefly our two systems CNN-LSTM and LSTM with attention. We have experimented both at character level and at word level. We have also explored using different pre-trained embeddings like Word2Vec, GloVe and also with embeddings learned from deep neural network models combined with several classifiers.

References

Itamar Arel, Derek C Rose, Thomas P Karnowski, et al. 2010. Deep machine learning-a new frontier in artificial intelligence research. *IEEE computational intelligence magazine*, 5(4):13–18.

Data	Presence of drug	Absence of drug
Train	3834	3572
Validation	959	893

Table 1: Task 1 Data Statistics

Data	personal medication intake	possible medication intake	non-intake
Train	2460	3932	5426
Validation	615	984	1357

Table 2: Task 2 Data Statistics

Method	Precision	Recall	F Score
LSTM-CNN with GloVe (word level)	0.8537	0.8537	0.8537
LSTMattention with GloVe and Naive Bayes classifier (word level)	0.8718	0.8718	0.8718
LSTM-CNN (character level)	0.8864	0.8864	0.8864

Table 3: Validation Data Results for Task 1

Method	Precision	Recall	F Score
LSTM-CNN with GloVe (word level)	0.8963	0.82433	0.85881
LSTMattention with GloVe and Naive bayes classifier) (word level)	0.86264	0.87202	0.86731
LSTM-CNN (character level)	0.91833	0.83976	0.87229

Table 4: Test Data Results for Task 1

Method	Micro-averaged Precision	Micro-averaged Recall	Micro-averaged F Score
LSTM-CNN with GloVe and GBDT (word level)	0.683	0.683	0.683
LSTM attention with Word2Vec) (word level)	0.706	0.694	0.694
LSTM-CNN (character level)	0.715	0.715	0.715

Table 5: Validation Data Results for Task 2

Dzmitry Bahdanau, Kyunghyun Cho, and Yoshua Bengio. 2014. Neural machine translation by jointly learning to align and translate. *arXiv preprint arXiv:1409.0473*.

William Chan, Navdeep Jaitly, Quoc Le, and Oriol Vinyals. 2016. Listen, attend and spell: A neural network for large vocabulary conversational speech recognition. In *Acoustics, Speech and Signal Processing (ICASSP), 2016 IEEE International Conference on*, pages 4960–4964. IEEE.

Method	Micro-averaged Precision	Micro-averaged Recall	Micro-averaged F Score
LSTM-CNN with GloVe and GBDT (word level)	0.350	0.365	0.358
LSTM attention with Word2Vec) (word level)	0.409	0.363	0.385
LSTM-CNN (character level)	0.408	0.407	0.408

Table 6: Test Data Results for Task 2

Michael Paul Graciela Gonzalez-Hernandez. Davy Weissenbacher, Abeed Sarker. 2018. Overview of the third social media mining for health (smm4h) shared tasks at emnlp 2018. In *Proceedings of the 2018 Conference on Empirical Methods in Natural Language Processing (EMNLP), 2018.*

Alex Graves. 2012. Supervised sequence labelling. In *Supervised sequence labelling with recurrent neural networks*, pages 5–13. Springer.

Yoon Kim. 2014. Convolutional neural networks for sentence classification. In *Proceedings of the 2014 Conference on Empirical Methods in Natural Language Processing (EMNLP)*, pages 1746–1751.

Yann LeCun, Yoshua Bengio, and Geoffrey Hinton. 2015. Deep learning. *nature*, 521(7553):436.

Kelvin Xu, Jimmy Ba, Ryan Kiros, Kyunghyun Cho, Aaron Courville, Ruslan Salakhudinov, Rich Zemel, and Yoshua Bengio. 2015. Show, attend and tell: Neural image caption generation with visual attention. In *International conference on machine learning*, pages 2048–2057.

Chunting Zhou, Chonglin Sun, Zhiyuan Liu, and Francis Lau. 2015. A c-lstm neural network for text classification. *arXiv preprint arXiv:1511.08630.*

Using PPM for Health Related Text Detection

Victoria Bobicev
Technical University of Moldova
victoria.bobicev@ia.utm.md

Victoria Lazu
Technical University of Moldova
victoria.lazu@ia.utm.md

Daniela Istrati
Technical University of Moldova
daniela.istrati@ia.utm.md

Abstract

This paper describes the participation of the LILU team in SMM4H challenge on social media mining for health related events description such as drug intakes or vaccinations.

1 The Tasks and the Data

The challenge included four tasks (Weissenbacher, 2018); we participated in Task 1: Automatic detection of posts mentioning a drug name — binary classification; and Task 4: Automatic detection of posts mentioning vaccination behavior — binary classification.

The data included medication-related posts on Twitter. The training data was available on the challenge site[1].

For the Task 1 the organizers provided 9624 annotated tweets' id numbers; 9130 tweets we downloaded using this data. The data was comparatively balanced: 4730 tweets that mention drug names and 4400 tweets that do not mention any drug or dietary supplement. The evaluation set consisted of 5384 twects.

For the Task 4 8180 annotated tweets' id numbers were provided. Only 6941 tweets we downloaded and the data was less balanced: 1979 tweets that mention influenza vaccination behavior and 4962 tweets that do not. The evaluation was performed on 161 tweets.

2 Method

We explored the PPM (Prediction by Partial Matching) model for automatic analysis of tweets. Prediction by partial matching (PPM) is an adaptive finite-context method for text compression that is a back-off smoothing technique for finite-order Markov models (Bratko et al., 2006). PPM produces a statistical language model which can be used in a probabilistic text classifier. Treating a text as a string of characters, a character-based PPM deals with different types of documents in a uniform way. PPM is based on conditional probabilities of the upcoming symbol given several previous symbols. A blending strategy for combining context predictions is to assign a weight to each context model, and then calculate the weighted sum of the probabilities:

$$P_{PPM}(x) = \sum_{i=1...m} \lambda_i p_i(x), \quad (1)$$

where $P_{PPM}(x)$ is the probability of the current character calculated using PPM method; $p_i(x)$ are conditional probabilities of this character on the base of the context of length i; λ_i are weights assigned to each conditional probability $p_i(x)$.

PPM is a special case of the general blending strategy. The PPM models use an escape mechanism to combine the predictions of all character contexts of length up to m, where m is the maximal length of the context; more details can be found in (Bobicev, 2007). The maximal length of a context equal to 5 in PPM model was proven to be optimal for text compression (Teahan, 1998) thus we used maximal length of a context equal to 5.

For example, the probability of character '*m*' in context of the word '*algorithm*' is calculated as a sum of conditional probabilities dependent on different context lengths up to the limited maximal length:

$$P_{PPM}('m') = \lambda_5 \cdot p('m' \mid 'orith') + \lambda_4 \cdot p('m' \mid 'rith') +$$
$$+ \lambda_3 \cdot p('m' \mid 'ith') + \lambda_2 \cdot p('m' \mid 'th') +$$
$$+ \lambda_1 \cdot p('m' \mid 'h') + \lambda_0 \cdot p('m') + \lambda_{-1} \cdot p('esc'),$$

Where λ_i is the normalization weight; 5 is the maximal length of the context; p('esc') is so

---

[1] https://healthlanguageprocessing.org/smm4h/social-media-mining-for-health-applications-smm4h-workshop-shared-task/

Proceedings of the 3rd Social Media Mining for Health Applications (SMM4H) Workshop & Shared Task, pages 65–66
Brussels, Belgium, October 31, 2018. ©2018 Association for Computational Linguistics

called 'escape' probability, the probability of an unknown character.

As a compression algorithm PPM is based on the notion of *entropy* introduced as a measure of a message uncertainty (Shannon, 1948).

Cross-entropy is the entropy calculated for a text if the probabilities of its characters have been estimated on another text (Teahan, 1998):

$$H_d^m = -\sum_{i=1}^{n} p^m(x_i) \log p^m(x_i) \qquad (2)$$

where   n is the number of symbols in a text d, H_d^m is the entropy of the text d obtained by model m, $p^m(x_i)$ is a probability of a symbol x_i in the text d obtained by model m.

The cross-entropy can be used as a measure for document similarity; the lower cross-entropy for two texts is, the more similar they are. Hence, if several statistical models had been created using documents that belong to different classes and cross-entropies are calculated for an unknown text on the basis of each model, the lowest value of cross-entropy indicates the class of the unknown text.

On the training step, we created PPM models for each class of posts; on the testing step, we evaluated cross-entropy of previously unseen posts using models for each class. Thus, cross-entropy was used as similarity metrics; the lowest value of cross-entropy indicated the class of the unknown posts.

PPM can be applied at the word level; however in most cases character level model better classify noisy texts with misspellings and slang (Bobicev, 2007).

3   Results

We performed a 10-fold cross-validation of the PPM based classification method on 6941 tweets, 1978 of which were from the positive class and 4963 from the negative class, and obtained: Precision = 0.839, Recall = 0.838, F-score = 0.839.

In order to improve the results we decided to remove less important words from the text before the model creation. The importance of words had been calculated using Gain Ratio (Quinlan, 1993):

$$GR = \frac{H(C) - \sum_{v \in V_i} P(v) * H(C|v)}{-\sum_{v \in V_i} P(v) \log P(v)} \qquad (3)$$

where *H(C)* is class entropy; *Vi* are features (in our case words); *v* are feature values (in our case 0 or 1; presence or absence of the word) and *P(v)* are probabilities of these values. Then, we removed a small number of words with the smallest Gain Ratio and repeated the experiment obtaining Precision = 0.861, Recall = 0.858, F-score = 0.859. The final result on the blind test set was as follows: Precision = 0.841, Recall = 0.860, F-score = 0.850. The mean result for all participating teams: P=0.890, R=0.872, F=0.880.

We proceeded in the same way for the task 4 and obtained: Precision = 0.842, Recall = 0.814, F-score = 0.828. The final result on the blind test set was as follows: Precision = 0.829, Recall = 0.808, F-score = 0.818. The mean result for all participated teams: P=0.826, R=0.858, F=0.840.

4   Conclusion

Our results are lower than the mean in both described tasks. The reasons of the low accuracy may be: (1) PPM is not suitable for this type of text classification; (2) more preprocessing of the texts should be done before classification phase; (3) all terms in text are treated uniformly; they can be weighted in some way while used in calculations. We plan to implement more sophisticated preprocessing and term weighting during next year challenge.

References

Bobicev, V. 2007 Comparison of Word-based and Letter-based Text Classification. *RANLP V, Bulgaria*, pp. 76–80.

Bratko A., Cormack G. V., Filipic B., Lynam T. R., Zupan B. 2006. Spam filtering using statistical data compression models, Journal of Machine Learning Research 7:2673–2698.

Quinlan, J.R. 1993. C4.5: Programs for Machine Learning. *Morgan Kaufmann, San Mateo, CA.*

Shannon, C. E. 1948. A Mathematical Theory of Communication. *The Bell System Technical Journal*, vol. 27, pp. 379–423, 623–656.

Teahan, W. 1998. Modelling English text, *PhD Thesis, University of Waikato, New Zealand.*

Weissenbacher, Davy, Abeed Sarker, Michael Paul, Graciela Gonzalez-Hernandez. 2018. Overview of the Third Social Media Mining for Health (SMM4H) Shared Tasks at EMNLP 2018, *in Proceedings of the 2018 Conference on Empirical Methods in Natural Language Processing (EMNLP), 2018.*

Leveraging Web Based Evidence Gathering for Drug Information Identification from Tweets

Rupsa Saha, Abir Naskar, Tirthankar Dasgupta and **Lipika Dey**

TCS Innovation Lab, India
(rupsa.s, abir.naskar, dasgupta.tirthankar, lipika.dey)@tcs.com

Abstract

In this paper, we have explored web-based evidence gathering and different linguistic features to automatically extract drug names from tweets and further classify such tweets into Adverse Drug Events or not. We have evaluated our proposed models with the dataset as released by the SMM4H workshop shared Task-1 and Task-3 respectively. Our evaluation results shows that the proposed model achieved good results, with Precision, Recall and F-scores of 78.5%, 88% and 82.9% respectively for Task1 and 33.2%, 54.7% and 41.3% for Task3.

1 Introduction

Use of data generated through social media for health studies is gradually increasing. It has been found that Adverse Drug Events (ADEs) are one of the leading causes of post-therapeutic death. Thus, their identification constitutes an important challenge. Social media platforms provide significant insights about drugs usage and their possible effects, as discussed by the general public outside the controlled environment of a trial program.

The shared task offers four different subtasks, out of which we focus on two : a) Sub Task 1 : Automatic detection of posts mentioning a drug name (binary classification) and b) Sub Task 3 : Automatic classification of adverse drug reaction mentioning posts (binary classification) (Weissenbacher et al., 2018). In the following section, we briefly describe the data used to build our systems. Section 3 describes the two systems in detail, followed by the results, and a final section consisting of our observations.

2 Data Description

2.1 Task 1

The provided training set of tweet ids and labels for Task 1 listed 9623 tweets, out of which 4975

Table 1: Number of tweets available, accessed and distribution of accessed tweets across Training and Validation

	Task-1			Task-3		
	Total	Label	#tweets	Total	Label	#tweets
Provided	9623	1 0	4975 4648	25623	1 0	2224 23399
Available	2496	1 0	1440 1056	13520	1 0	1109 12411
Train	2121	1 0	1219 902	10817	1 0	888 9929
Validation	375	1 0	221 154	2703	1 0	221 2482
Test	5382			5000		

were marked with label 1 ("yes"), i.e. tweets containing mention of drug product names and/or dietary supplements. However, due to network constraints or unavailability of tweets, we could only obtain 2496 tweets. Of these, 1440 were of label "1", and 1056 were labeled "0". For the purpose of building our system, we split this set in a 85:15 ratio, and 375 tweets (216-"1", 159-"0") were used for validation, while the rest were used for building the system.

For Task 3, the provided training set of tweet ids and labels listed 25623 tweets out of which we were able access 13520. Out of these, a mere 1109 were labeled as tweets containing mention of adverse drug events. We divided this set in a 80:20 ratio, with 2703 tweets (221- "1", 2482 - "0") as the validation set and the remaining 10817 for training.

3 System Description

3.1 Task 1 : Automatic detection of posts mentioning a drug name

As preprocessing, each tweet is tokenized and tagged using the Ark-Tweet-NLP tool (Owoputi et al., 2012). From the resultant tokens, we considered those that are tagged as Proper Nouns or Common Nouns. Such tokens are passed to the Information Gathering Module.

Proceedings of the 3rd Social Media Mining for Health Applications (SMM4H) Workshop & Shared Task, pages 67–69
Brussels, Belgium, October 31, 2018. ©2018 Association for Computational Linguistics

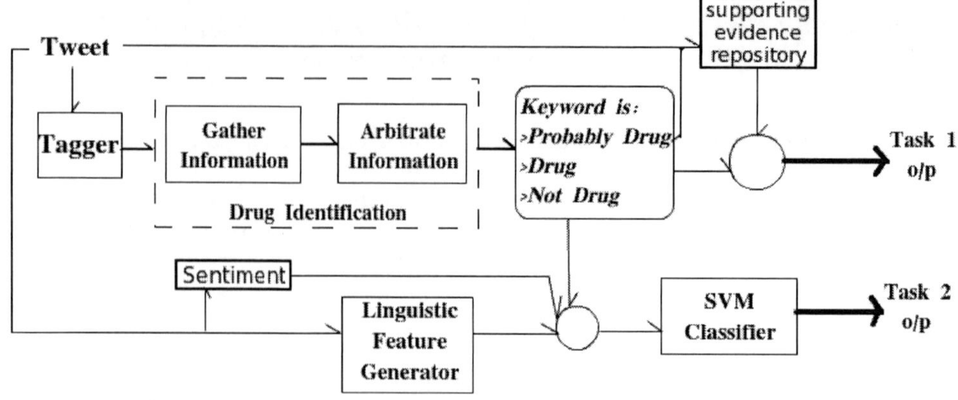

Figure 1: Overview of architecture

Table 2: Preliminary results from internal evaluation on the training dataset

Approach	Class	Precision	Recall	F-Score	Accuracy
	Class 0	0.58	0.70	0.63	
Task 1	Class 1	0.70	0.57	0.63	0.63
	Wt. Avg	0.64	0.63	0.63	
	Class 0	0.96	0.8	0.88	
Task 3 (1)	Class 1	0.23	0.64	0.34	0.79
	Wt. Avg	0.89	0.79	0.83	
	Class 0	0.95	0.85	0.9	
Task 3 (2)	Class 1	0.27	0.58	0.37	0.83
	Wt. Avg	0.9	0.83	0.86	

This module retrieves information relevant to the keyword from three different sources : Wikidata (Vrandečić and Krötzsch, 2014), Wikipedia data dumps (Wu and Weld, 2010) and Wordnet (University, 2010). The module searches for evidence that a word represents a drug/supplement, in the corresponding gloss, hierarchy structure, and web page structure, as obtained from each source. Wikidata is the source of structured information, and presence or absence certain keys (e.g. RxNorm Id.,drug interaction etc.) are used as evidence. On the other hand, Wikipedia is mostly unstructured textual information. From this source, evidence may be found in the form of the definition of the keyword, the presence of "side effects" of the keyword, the hierarchical category the entry belongs to, among other ways. From Wordnet, we use both the gloss and the hierarchy structure.

The obtained information is further fed to the Information Arbitration Module. The Arbitration module considers the different information obtained with regard to a particular keyword, and returns a judgment as to whether the the keyword is "Not Drug", "Probably Drug " or "Drug".

In case a keyword receives a "Probably Drug" judgment, it may be a drug name depending on the information obtained from neighbouring tokens. In such cases, we extract the most frequent keywords co-occurring with the keywords to create a repository of terms. this supporting evidence repository contains a collection of patterns, obtained from the training set, which dictate under which conditions a "Probably Drug" keyword can be upgraded to a "Drug". e.g. "Protein" by itself is not a supplement name, however, "Protein shakes" is, when used as treatment.

3.2 Task 3 : Automatic classification of adverse drug reaction mentioning posts

For classification, we employ a SVM based classifier with a polynomial kernel (Dasgupta et al., 2017). The features used are: *(a) PMI*: the Pointwise Mutual Information (PMI)(Bouma, 2009) between all possible bigram pairs are considered. Co-occurrence is counted at the sentence level, i.e. *P(i, j)* is estimated by the number of sentences that contain both terms W_i and W_j , and *P(i)* and *P(j)* are estimated by counting the total sentences containing W_i and W_j , respectively. Only those bi-grams whose PMI score exceeds the *average + stddev* threshold, are retained as features, *(b) Term Relevance*: all unigram terms that are relevant to

Table 3: Results of Experiments on Final Test Set

Task		Precision	Recall	F-Score
Task 1	Team ART	0.79	**0.88**	0.83
	Task Average	**0.89**	0.87	**0.88**
Task 3	Team ART (1)	0.305	**0.627**	0.411
	Team ART (2)	0.332	0.547	**0.413**
	Task Average	**0.39**	0.52	0.40

the positive class, *(c) Dependency feature counts*: counts of all Stanford typed dependency features, and *(d) Drug Name*: The drug names present in the tweet, as obtained employing the same Drug Identification Module mentioned in 3.1. Since the data is heavily skewed in favour of negative examples, we train a total of 11 models, each with a non-skewed subset of the data. The training data for each model consists of 909 positive and approximately 925 negative examples, with negative examples, sampled randomly. For each test data, each of the 11 models predict the "yes"/"no" label, and all the predictions are fed to an arbitrator for a final decision.

As an enhancement, we also use the sentiment polarity score as an additional feature, using the VADER sentiment analysis tool (Gilbert, 2014). Using sentiment does result in a performance improvement, as noted in section 4.

4 Results and Observations

The results as obtained from splitting the initial data into training and test sets are tabulated in Table 2. For Task 1, we report the Precision, Recall, F-Score and Accuracy on the whole of the training set. However, for Task 3, since the data is skewed, we report all three versions (for negative class, for positive class and for weighted average of both classes) of these same parameters. Task 3 (1) represents the results for experimentation without using sentiment polarity, and Task 3 (2) are experiments with the sentiment factor.

The results on the final test set are reported in Table 3. We compare our result with the mean of results obtained by other participating teams. For Task 3, only the results with respect to the positive class was from others were available.

The results for Task 1 are poor because, while the Drug Identification module is successful in pointing out keywords which are drugs/supplements in the tweet, it does not have the capability to distinguish whether that keyword

is used to imply medication or not. For example, in *"I wanna name my first child vyvanse"*, while *"vyvanse"* is a drug, here it is clearly not used in a medication sense of the term, and the given label is 0. Our method fails in such cases.

Results for Task 3 may benefit on using a more robust sentiment feature scorer, especially one that is trained on drug tweets themselves. We can also use different classification methods to test if results improve further.

References

Gerlof Bouma. 2009. Normalized (pointwise) mutual information in collocation extraction. *Proceedings of GSCL*, pages 31–40.

T Dasgupta, A Naskar, and L Dey. 2017. Exploring linguistic and graph based features for the automatic classification and extraction of adverse drug effects. In *18th International Conference on Computational Linguistics and Intelligent Text Processing*.

CJ Hutto Eric Gilbert. 2014. Vader: A parsimonious rule-based model for sentiment analysis of social media text. In *Eighth International Conference on Weblogs and Social Media (ICWSM-14)*. *Available at (20/04/16) http://comp. social. gatech. edu/papers/icwsm14. vader. hutto. pdf*.

Olutobi Owoputi, Brendan OConnor, Chris Dyer, Kevin Gimpel, and Nathan Schneider. 2012. Part-of-speech tagging for twitter: Word clusters and other advances. *School of Computer Science*.

Princeton University. 2010. Wordnet. *About WordNet*.

Denny Vrandečić and Markus Krötzsch. 2014. Wikidata: A free collaborative knowledgebase. *Commun. ACM*, 57(10):78–85.

D Weissenbacher, A Sarker, M Paul, and G Gonzalez-Hernandez. 2018. Overview of the third social media mining for health (smm4h) shared tasks at emnlp 2018. In *Proceedings of the 2018 Conference on Empirical Methods in Natural Language Processing*.

Fei Wu and Daniel S Weld. 2010. Open information extraction using wikipedia. In *Proceedings of the 48th annual meeting of the association for computational linguistics*, pages 118–127. Association for Computational Linguistics.

CLaC at SMM4H Task 1, 2, and 4

Parsa Bagherzadeh, Nadia Sheikh, Sabine Bergler

CLaC Labs
Department of Computer Science and Software Engineering
Concordia University, Montreal
{p_bagher, n_she, bergler} @ encs.concordia.ca

Abstract

CLaC Labs participated in Tasks 1, 2, and 4 using the same base architecture for all tasks with various parameter variations. This was our first exploration of this data and the SMM4H Tasks, thus a unified system was useful to compare the behavior of our architecture over the different datasets and how they interact with different linguistic features.

1 Base system

The base system is a feed-forward neural network with a recurrent neuron. We decided to explore that architecture for independent purposes and used the SMM4H tasks to compare performance on different datasets and task descriptions.

We considered three variations of this architecture:

Full: A recurrent neuron that outputs a 20 dimensional vector is followed by a 3 layer feedforward neural net, all embedded in two decision neurons with soft-max activations. The feedforward network has 50, 25 and 12 neurons in first, second and third layers respectively. Unless otherwise mentioned, the network has been trained for 100 epochs.

The recurrent neuron consists of an LSTM cell using $tanh$ activations [Hochreiter and Schmidhuber, 1997]. The activation functions for the feedforward networks are also $tanh$.

NR: Only the recurrent neuron and the decision neurons are used, the feed-forward (N)etwork is (R)emoved.

Full+At: Attention is added to the full architecture. In contrast to Full, where the LSTM cell outputs a single vector, in Full+At, the recurrent neuron outputs the sequence of each time step.

We used the Keras package [Chollet and others, 2015] to implement the neural networks using TensorFlow as backend [Abadi *et al.*, 2015].

1.1 Input parameters

Tweets are normalized to a size of 25, padded with leading zeros or shortened from the end as required.

The input per tweet consists thus of 25 word vectors of size 100 compiled by the Word2Vec method [Mikolov *et al.*, 2013] over the training data. The Gensim package [Řehůřek and Sojka, 2010] is used for the training of word vectors. The minimum number of occurrences for a word to be considered in the vocabulary is 1 and the window size has been set to 5. Other parameters involved in word vector training were left to the default values of the Gensim package.

Tweet representations are then binned to a batch size of 5, unless otherwise indicated.

2 Text features and knowledge sources

We also experimented with a few linguistic text features and a gazetteer list to see whether they might influence the results.

2.1 Gazetteer

Inspired by Task 1, detection of drug mentions, we scraped the *name* field of *product* fields in Drug-Bank [Wishart *et al.*, 2017] to compile a gazetteer list for drugs. Due to time constraints, this resource was only minimally refined and contained many multi-word drug names such as *One A Day* and dosage specifications (*Aspirin 80mg*). The gazetteer information was appended to the word vector. Runs that use the gazetteer are identified as *+Gaz*.

Proceedings of the 3rd Social Media Mining for Health Applications (SMM4H) Workshop & Shared Task, pages 70–73
Brussels, Belgium, October 31, 2018. ©2018 Association for Computational Linguistics

Table 1: Training and validation results for Task 1

Architecture	WV trained on Task 1 data					WV trained on Task 1 +Task 2 data				
	Train Acc	Valid. Acc	Precision	Recall	F1	Train Acc	Valid. Acc	Precision	Recall	F1
Full	0.76	0.55	0.87	0.54	0.66	0.95	0.64	0.87	0.66	0.75
Full+Gaz	0.82	0.56	0.90	0.52	0.66	0.96	0.60	0.88	0.60	0.71
Full+POS	0.75	0.59	0.89	0.57	0.70	0.93	0.59	0.89	0.57	0.70
Full+Modality	0.76	0.55	0.87	0.53	0.66	0.94	0.60	0.87	0.61	0.72
Full+Gaz+POS	0.82	0.57	0.89	0.54	0.67	0.93	0.64	0.88	0.64	0.74
Full+Gaz+POS+Mod	0.81	0.50	0.90	0.42	0.58	0.96	0.62	0.88	0.64	0.73
NR	0.74	0.64	0.87	0.65	0.75	0.94	0.61	0.87	0.62	0.72
NR+Gaz	0.82	0.53	0.88	0.50	0.63	0.96	0.60	0.88	0.60	0.71
NR+POS	0.74	0.57	0.86	0.57	0.69	0.92	0.59	0.88	0.64	0.74
NR+Gaz+POS	0.81	0.59	0.87	0.59	0.70	0.94	0.63	0.87	0.63	0.73
Full+All	0.82	0.64	0.85	0.69	0.76	0.95	0.64	0.90	0.63	0.74
Full+All+At	0.85	0.65	0.86	0.70	0.77	0.95	0.65	0.91	0.68	0.78

2.2 Linguistic features

We used a CLaC pipeline in the GATE environment to extract linguistic features for each tweet. Third party processing resources in our pipeline include the ANNIE Twitter Tokenizer [Cunningham *et al.*, 2002], the Hashtag Tokenizer [Maynard and Greenwood, 2014] and the Stanford Part-Of-Speech Tagger with a model trained on tweets [Toutanova *et al.*, 2003].

Following sentence splitting, tweets were tokenized and Twitter specific tokens (@name and URLS) were removed from the token set. The remaining tokens were assigned one of 36 part-of-speech tags, resulting in a feature value range of integers from 1 to 36.

Following [Doandes, 2003], the part-of-speech tags were used to identify verb clusters. Voice, tense and aspect were assigned to each verb cluster, and the main verb in each verb cluster was identified. These features were also added to the respective word vectors of the main verbs.

We selected only indicative tenses for our binary *tense* feature.

Tokens were also checked against two ad hoc gazetteer lists of explicit negation triggers and modality terms and the binary features *neg* and *mod* were added to the respective word vectors.

Thus we created 4 linguistic features (*tense, voice, POS, and modality*) in addition to the gazetteer feature, that can be appended to word vectors for those words onto which the features project.

3 Task 1

Task 1 was a basic binary categorization task, identifying tweets where a drug was mentioned in its medical sense (the detailed description of the tasks and data can be found in the overview paper [Weissenbacher *et al.*, 2018]). The training data

consisted of over 9000 tweets, balanced in both categories.

Table 1 shows the results from some of the runs we compared in order to evaluate the effectiveness of our features. We selected a validation set of 1000 tweets from the training data and trained on the remaining tweets. We compared the training accuracy and the validation accuracy to get some indication of the degree of overtraining. We observe that the difference for training accuracy and validation accuracy is surprisingly small for such a small dataset. Moreover, the differences between our different feature bundles is also rather small. The gazetteer list led to a marked improvement for training accuracy, but not necessarily validation accuracy. Paradoxically, the two best validation accuracy performances came from NR and Full+All (with Full+All+At adding a percentage point). That means that on the validation data, the contribution of the neural net plus gazetteer plus all linguistic feature (plus attention) was matched by simply removing the neural net (NR).

We achieved a greater performance increase in training accuracy across all our configurations when training on Task 2 training data as well as on Task 1 training data. This improvement carries over to validation accuracy and F1 measure, but inconsistently. However, the overall results of different configurations showed less variation when also training on task 2 training data. We speculate that this stabilization may stem from some disruptive effect of data from another task (but that can be expected to contain drug mentions) which might counterbalance overfitting. Our competition runs were all trained on both, Task 1 and Task 2 training data.

It was clear from the beginning that our architecture is severely mismatched to the simple categorization task. The very small difference that our different experiments generated show that the variations do not truly access different tweets. The extremely high training accuracy indicates to us a high degree of overfitting, with the danger of making the entire system somewhat brittle. Table 2 shows that our best competition run on Task 1 was with the Full architecture, the addition of the gazetteer list and two linguistic features reduced the performance. But the near equal performance of Full and NR+Gaz+POS[1] confirms the findings of the validation data, namely that the performance contribution of the network can be matched by the gazetteer list plus some linguistic features. Interestingly, our official test results top the results we obtained on our validation set, which shows that the performance in this case was stable. The performance difference between the best and the last system was 0.1399.

Table 2: Official Task 1 results for CLaC Difference between best and last system score for this task was 0.1399

	P	R	F
NR+Gaz+POS	0.75	0.80	0.77
Full	0.79	0.77	0.78
Full+Gaz+Mod+POS	0.76	0.76	0.76
Competition Mean	0.89	0.87	0.88

4 Task 2

Task 2 had a semantic component that Task 1 lacked: it concerned distinguishing actual medication intake from possible medication intake and mere mention of a medication in a 3-way decision. We augmented the basic architecture with a third decision neuron for this task.

The training data size for Task 2 was 14482 tweets that were highly imbalanced. Again, a validation set of 1000 tweets was randomly selected from the training data.

Table 3 shows that the richer task definition led to a greater variance in team performance: the difference between the first and last placed team's best runs is .341 micro-averaged F measure. Unlike for Task 1, our performance was not commensurate with our validation performance: in validation runs Full+All+At was also the best run with a validation accuracy of 0.85. Note, that our performance is determined in part by the lowest recall.

These results suggest to us that firstly, a custom tailored architecture that better addresses the task can make a greater difference and that our architecture showed more signs of overfitting than in Task 1.

Table 3: Official Task 2 micro-averaged results

Team	P	R	F
UChicagoCompLx	0.654	0.783	0.713
Light_task2	0.520	0.491	0.505
Tub-Oslo-task2-predictions	0.478	0.458	0.468
IRISA_team_task2	0.434	0.501	0.465
IIT_KGP	0.408	0.407	0.408
UZH	0.371	0.437	0.401
CLaC Full+All+At	0.402	0.366	0.383
Techno	0.327	0.432	0.372

5 Task 4

Task 4 was the most semantics oriented task we attempted. The binary task was to identify tweets that clearly indicate that someone received, or intended to receive, a flu vaccine.

Of the 8000 tweets mentioned in the task description, only 4502 tweets could be downloaded for our training data. Despite the very small size of the training data and the potentially deeper semantic distinction, our system performed the closest to the competition mean. Note that the general drug name gazetteer list was not useful for this task.

Table 4: Official Task 4 results for CLaC.

	P	R	F1
Full+All	0.70	0.89	0.78
NR	0.76	0.46	0.57
Full+Voice+Tense	0.75	0.65	0.69
Competition Mean	0.82	0.85	0.84

CLaC's best run was by Full+All. It is interesting that what appeared to us as the semantically most difficult task has our best performance (measured in distance to the competition mean) due to a recall of .89. We speculate that there may be certain linguistic patterns that our features were able to detect that made this task more amenable to our architecture (in comparison) based on the fact that Full+All outperforms NR and Full+Voice+Tense significantly.

---

[1] less obvious due to rounding in Table 2

6 Conclusion

CLaC decided late to participate in SMM4H with a uniform architecture to test across several tasks that was not inspired by them. Our conclusion is that the architecture and in particular the input binning and normalizing techniques have to be carefully reviewed, as they risk ignoring key terms in the input. The linguistic features showed some effect, as did the addition and removal of the network. Repeatedly, trial runs showed that removing the network could be offset by adding linguistic features to the recurrent neuron. The detailed interplay of these parameters has to be further studied.

However, we conclude that using the same architecture across several tasks (that are related, but differ significantly) is an interesting exercise and allowed us to gain additional insight. Despite its potential for gross overfitting, the architecture has shown promise. The linguistic features also proved effective, and most importantly, the two components interplay effectively as demonstrated in the fact that in two tasks Full+All was our best performing run.

While each of the three tasks is interesting in itself and clearly has relevance to society at large, we find the juxtaposition of the three tasks very interesting for the ML/NLP researcher.

References

Martín Abadi, Ashish Agarwal, Paul Barham, Eugene Brevdo, Zhifeng Chen, Craig Citro, Greg S. Corrado, Andy Davis, Jeffrey Dean, Matthieu Devin, Sanjay Ghemawat, Ian Goodfellow, Andrew Harp, Geoffrey Irving, Michael Isard, Yangqing Jia, Rafal Jozefowicz, Lukasz Kaiser, Manjunath Kudlur, Josh Levenberg, Dandelion Mané, Rajat Monga, Sherry Moore, Derek Murray, Chris Olah, Mike Schuster, Jonathon Shlens, Benoit Steiner, Ilya Sutskever, Kunal Talwar, Paul Tucker, Vincent Vanhoucke, Vijay Vasudevan, Fernanda Viégas, Oriol Vinyals, Pete Warden, Martin Wattenberg, Martin Wicke, Yuan Yu, and Xiaoqiang Zheng. TensorFlow: Large-scale machine learning on heterogeneous systems, 2015. Software available from tensorflow.org.

François Chollet et al. Keras. https://keras.io, 2015.

H. Cunningham, D. Maynard, K. Bontcheva, and V. Tablan. GATE: A framework and graphical development environment for robust NLP tools and applications. In *Proceedings of the 40th Anniversary Meeting of the Association for Computational Linguistics (ACL'02)*, Philadelphia, 2002.

M. Doandes. Profiling for belief acquisition from reported speech. Master's thesis, Concordia University, 2003.

Sepp Hochreiter and Jürgen Schmidhuber. Long short-term memory. *Neural computation*, 9(8):1735–1780, 1997.

DG Maynard and Mark A Greenwood. Who cares about sarcastic tweets? investigating the impact of sarcasm on sentiment analysis. In *LREC 2014 Proceedings*. ELRA, 2014.

Tomas Mikolov, Ilya Sutskever, Kai Chen, Greg S Corrado, and Jeff Dean. Distributed representations of words and phrases and their compositionality. In *Advances in neural information processing systems*, pages 3111–3119, 2013.

Radim Řehůřek and Petr Sojka. Software Framework for Topic Modelling with Large Corpora. In *Proceedings of the LREC 2010 Workshop on New Challenges for NLP Frameworks*, pages 45–50, Valletta, Malta, May 2010. ELRA. http://is.muni.cz/publication/884893/en.

Kristina Toutanova, Dan Klein, Christopher D Manning, and Yoram Singer. Feature-rich part-of-speech tagging with a cyclic dependency network. In *Proceedings of the 2003 Conference of the North American Chapter of the Association for Computational Linguistics on Human Language Technology-Volume 1*, pages 173–180. Association for Computational Linguistics, 2003.

Davy Weissenbacher, Abeed Sarker, Michael Paul, and Graciela Gonzalez-Hernandez. Overview of the third social media mining for health (smm4h) shared tasks at emnlp 2018. In *Proceedings of the 2018 Conference on Empirical Methods in Natural Language Processing (EMNLP)*, 2018.

DS. Wishart, YD. Feunang, AC. Guo, EJ. Lo, A. Marcu, JR. Grant, T. Sajed, D. Johnson, C. Li, Z. Sayeeda, N. Assempour, I. Iynkkaran, Y. Liu, A. Maciejewski, N. Gale, A. Wilson, L. Chin, R. Cummings, D. Le, A. Pon, C. Knox, and M. Wilson. Drugbank 5.0: a major update to the drugbank database for 2018. *Nucleic Acids Res.*, 2017.